Biomass Based Products

Edited by Eduardo Jacob-Lopes, Leila Queiroz Zepka and Rosangela Rodrigues Dias

Published in London, United Kingdom

Biomass Based Products
http://dx.doi.org/10.5772/intechopen.1003405
Edited by Eduardo Jacob-Lopes, Leila Queiroz Zepka and Rosangela Rodrigues Dias

Contributors
Abel Olajide Olorunnisola, Adilson Gonçalves, Ana Amorim, Anabela Romano, Bruno Medronho, Catarina Fernandes, Christian Krüger, Fernando Masarin, Fernando Paz-Cedeno, Hugo Duarte, Inês Mansinhos, Isabela dos Anjos, Ivana Krkljuš, João Brás, Jun Sun, Lethiwe D. Mthembu, Luís Alves, Maria José Aliaño-González, Md. Zakir Hossain, Mohammad Ali Sabbaghi, Muhammad Badrul Islam Chowdhury, Musiliu Ademuiwa Onilude, Naila Mori, Nirmala Deenadayalu, Peter Saling, Ping Hu, Raphael Segun Bello, Raquel Rodríguez-Solana, Ruochen Wu, Samuel Oliveira, Sandra Gonçalves, Solange Magalhães, Temidayo Emmanuel Omoniyi, Weijin Wang, Zhaofeng Wu

First published in London, United Kingdom, 2025 by IntechOpen
IntechOpen is the global imprint of INTECHOPEN LIMITED, registered in England and Wales, registration number: 11086078, 167-169 Great Portland Street, London, W1W 5PF, United Kingdom

For EU product safety concerns: IN TECH d.o.o., Prolaz Marije Krucifikse Kozulić 3, 51000 Rijeka, Croatia, info@intechopen.com or visit our website at intechopen.com.

British Library Cataloguing-in-Publication Data
A catalogue record for this book is available from the British Library

Biomass Based Products
Edited by Eduardo Jacob-Lopes, Leila Queiroz Zepka and Rosangela Rodrigues Dias
p. cm.
Print ISBN 978-1-83634-305-9
Online ISBN 978-1-83634-304-2
eBook (PDF) ISBN 978-1-83634-306-6

If disposing of this product, please recycle the paper responsibly.

Meet the editors

Eduardo Jacob-Lopes is an associate professor at the Department of Food Technology and Science at the Federal University of Santa Maria in Brazil. His research focuses on a multidisciplinary view of applied phycology and its intersection with bioprocess engineering and sustainable engineering.

Leila Queiroz Zepka is an associate professor at the Department of Food Technology and Science at the Federal University of Santa Maria in Brazil. Her research covers issues related to microalgae with an emphasis on food science, natural pigments, and novel healthy foods.

Rosangela Rodrigues Dias is a researcher at the Department of Food Technology and Science at the Federal University of Santa Maria in Brazil. His research focuses on the life cycle assessment of bio-based processes and products.

Contents

Preface

Biomass is the future. The origin of the acclaimed use of biomass for a net-zero economy lies in its apparent sustainability and renewability. It is now categorically clear that moving away from fossil resources is necessary. There is no easy solution, and biomass is not a magic solution, but its intersection with the environment is a crucial aspect of creating a sustainable economy. Under this horizon, as part of a future being written, in this book, we explore biomass resources that can be used for applications such as energy, chemicals and materials, and bioremediation, as well as biomass processing technologies and operational, economic, social, political, and regulatory challenges that haunt the environmentally beneficial, socially responsible, and economically viable use of biomass. This book brings together contributions from renowned specialists who launch a close look at the state-of-the-art of biomass and biomass-based products.

Eduardo Jacob-Lopes, Leila Queiroz Zepka and Rosangela Rodrigues Dias
Federal University of Santa Maria,
Santa Maria, Brazil

Chapter 1

Biomass Conversion Technologies: Transforming Organic Matter into Energy and Materials

Ruochen Wu

Abstract

This chapter provides a comprehensive exploration of the various technologies used to convert biomass into valuable products, such as biofuels, biochemicals, and bioenergy. It thoroughly examines the three main categories of current biomass conversion technologies: thermochemical, biochemical, and physicochemical processes. Thermochemical conversion includes processes such as combustion, gasification, and pyrolysis, which utilize heat to transform biomass. Biochemical conversion involves biological processes such as anaerobic digestion and fermentation to produce energy and chemicals. Physicochemical conversion, such as transesterification, chemically alters biomass to create bio-based products. The chapter meticulously examines the principles, mechanisms, and applications of each technology, highlighting their role in creating sustainable, renewable energy solutions and contributing to waste management and environmental protection.

Keywords: thermochemical conversion, biochemical conversion, physicochemical conversion, biofuel, bioproducts

1. Introduction

The growing interest in renewable energy stems from two main factors: (1) heightened concern over global climate change and the environmental effects of fossil fuels, and (2) rising worries about the security and sustainability of fossil fuel supplies [1]. Interest in renewable energy arises from two sources: (1) increasing concern about global climate change and the environmental impact of fossil energy sources and (2) increasing anxiety regarding the security and longevity of fossil fuel [1]. Approximately 14% of the world's primary energy comes from biomass [2]. In developing countries, the percentage is notably impacted, as biomass is mainly burned for household heating and cooking. In contrast, biomass accounts for less than 3% of primary energy in industrialized nations. Globally, only around 40% of the biofuel potential is utilized, with Asia being the only region where consumption surpasses the available potential [2]. This highlights the significant potential for expanding biomass usage as a sustainable alternative or complement to coal. A strategy proposed by the US Department of Agriculture and Energy suggests that biomass sourced from forest and agricultural lands could replace up to 30% of the current petroleum consumption in the US by 2030 [1, 2].

IntechOpen

As the world increasingly shifts toward sustainable and renewable energy sources, biomass has emerged as a key player in the production of biofuels, biochemicals, and bioenergy [3, 4]. In order to meet the Net Zero Emissions by 2050 Scenario, biomass, as a zero-carbon fuel, is essential to advancing sustainable energy on a global scale [5]. Biomass, which includes organic material from plants, animals, and waste, can be converted into a variety of valuable products using different conversion technologies [6]. These technologies not only decrease our dependence on fossil fuels but also offer solutions for waste management and the mitigation of greenhouse gas emissions [7].

Biomass conversion technologies are broadly categorized into three types: thermochemical, biochemical, and physicochemical processes [8]. Each of these methods leverages different scientific principles and mechanisms to unlock the energy and material potential stored in biomass. Thermochemical processes, for instance, use heat to break down biomass into energy-rich gases, liquids, or solids, while biochemical processes employ microorganisms or enzymes to convert biomass into biofuels and other high-value chemicals. The physicochemical process involves both physical and chemical reactions that modify biomass into products, such as biodiesel. The thermochemical process happens relatively rapidly at high temperatures that are a few 100°C and sometimes up to over 1000°C. In contrast, the biochemical process proceeds at a few tens of degrees Celsius above ambient temperature, thus it can be quite slow even in the presence of catalysts. The physiochemical process often combines physical actions, such as changes in phase (solid, liquid, gas) or mixing, with chemical reactions that alter the molecular structure or composition of a substance. Therefore, the operating conditions required for physicochemical processes can vary widely depending on the specific process and materials [9–20].

This chapter will delve into these technologies in depth, examining the underlying principles, major processes, and their respective applications. By understanding these conversion pathways, we gain insight into how biomass can be effectively transformed into clean, sustainable energy and materials, helping address critical environmental and energy challenges. As we move forward, these technologies play a pivotal role in the global transition to a circular bioeconomy, where waste is reduced, and renewable resources are maximized.

2. Biomass conversion technologies

Biomass represents the organic material that comes from plants and animals. Plants produce biomass through photosynthesis, which stores the chemical energy. Biomass may contribute little or no net atmospheric CO_2 when harvested sustainably and used as a fuel. Biomass sources for energy include: (1) wood and wood processing wastes, such as firewood, wood pellets, and wood chips, and black liquor from pulp and paper mills; (2) agricultural crops and waste materials, such as corn, soybeans, sugar cane, switchgrass, woody plants, and algae, and crop and food processing residues; (3) biogenic materials in municipal solid waste, such as paper, cotton, and wool products, and food, yard, and wood wastes; (4) animal manure and human sewage for producing biogas/renewable natural gas (**Figure 1**) [21–24].

2.1 Thermochemical conversion

Thermochemical conversion is a technique that occurs at high temperatures to decompose biomass into energy and valuable products such as biofuels, syngas, and

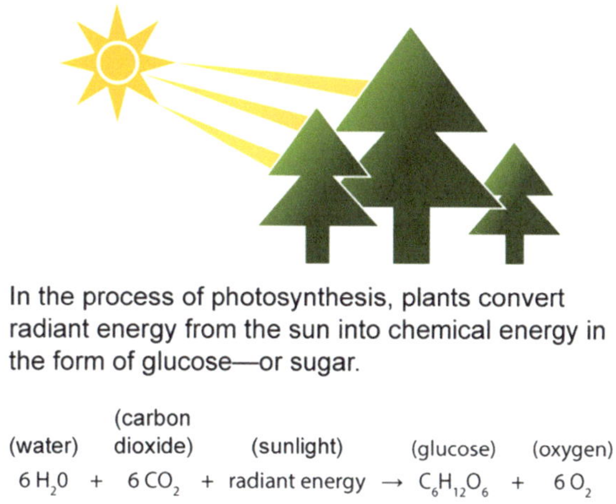

Figure 1.
Scheme of the photosynthesis process.

biochar. Syngas is a mixture of primarily CO and H_2 gas components with varying amounts of CO_2 and H_2O. The syngas converts to other products in one of a variety of catalytic, pressurized processes to produce base chemicals, transportation fuels, or finished chemicals. Several typical applications for syngas in the industry are water-gas shift, catalytic synthesis, fermentation, etc. By utilizing high temperatures, thermochemical conversion can transform organic materials into various energy-rich forms, which makes thermochemical conversion one of the most efficient pathways for extracting energy from biomass. This approach is particularly effective for large-scale applications, including power generation, heat production, and the creation of industrial feedstocks. The three main thermochemical conversion processes are categorized as combustion, gasification, and pyrolysis, which are determined by the temperature range for each stage. **Figure 2** indicates that at 0–250°C, moisture in biomass will evaporate. As the temperature goes up to 250–500°C (sometimes can be up to 700°C), pyrolysis (or devolatilization) becomes the dominant stage, which consists of involves many complex reactions and product gases. Combustion and gasification require similar reacting temperatures (typically 800–1100°C), while the key differences are the degree of oxidation and heating rate [25].

Direct combustion is the most common and simplest foundation approach, where biomass is burned in the presence of oxygen to generate heat, which can then be used directly, or converted into electricity. This process is widely used in biomass power plants and combined heat and power (CHP) systems to produce renewable energy while mitigating carbon emissions. The operation procedure is achieved by rapid oxidation reactions of fuel (biomass raw materials) and oxygen to obtain thermal energy and flue gas, consisting mainly of CO_2 and H_2O. Several disadvantages hinder the promising application of direction combustion for biomass energy utilization. For instance, high-moisture amounts in the biomass raw materials, agglomeration, and fly-ash deposition due to alkali compounds in biomass, along with the high cost of transporting bulky biomass are currently the most challenging limitations for direction combustion technology [9, 12].

Gasification is the conversion of biomass at elevated temperatures in a limited-oxygen environment to produce syngas, a mixture of carbon monoxide, hydrogen, methane, nitrogen, carbon dioxide, and some hydrocarbons. Syngas can be used

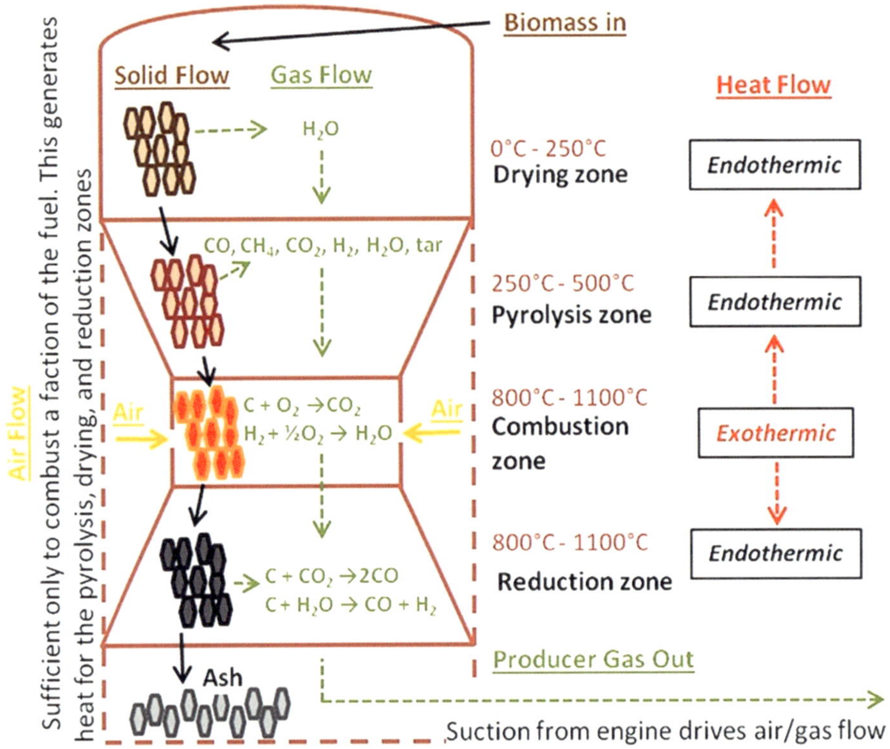

Figure 2.
Temperature zones of drying, pyrolysis, combustion, and gasification (reduction) stages [25].

directly in gas turbines for electric power generation or be further processed to create synthetic fuels using catalysts or microorganisms (syngas fermentation). The carbonaceous biomass solids or liquids can be gasified into smaller and lower molecular-weight gas mixtures, similar to coal gasification. Coal gasification has been under development for about 200 years, beginning with the aim of providing heating and lighting. Coal gasification has a long and rich history of research in the modern Western world, but its track record of successful operation is much shorter and less impressive. However, countries such as Nazi Germany and Apartheid-era South Africa, which were cut off from petroleum supplies, as well as nations with limited access to oil like China and India, have effectively used coal gasifiers to make significant contributions to their fuel and chemical supplies. These operational processes, along with a few successful government-led gasification demonstrations, make up the current fleet of gasifiers in current operation. Due to the high volatile content in biomass raw materials, biomass gasification can occur more readily than that of coal. This technology possesses higher energy efficiency and flexibility than direct combustion, and produces fewer emissions, making it an attractive option for sustainable energy production, hydrogen production, and synthesis of fuels and high-value chemicals. As petroleum and natural gas become more expensive with reduced reservations in the world, biomass gasification represents promising applications in the near future [1, 9]. The gasification stage in these applications accounts for more than 75% of the total capital cost and poses an even higher proportion of the technical risk. Gasifiers pose significant operational and design difficulties, and even when fully implemented, they tend to be the most problematic and inelegant portions of the overall process [26, 27]. In addition, sizeable energy input, tar production, and significant energy loss especially in smaller-scale operations and plants are the prominent factors that make biomass gasification less ideal in some cases, though ongoing technological advancements aim to address these issues.

Pyrolysis takes place in the absence or very limited presence of oxygen and breaks down biomass into mainly bio-oil (liquid), biochar (carbo-rich solid), and syngas (gas) at moderately high temperatures (250–700°C).

Pyrolysis is a highly versatile process, with bio-oil serving as a precursor for biofuels, biochar used to enhance soil, and syngas providing an energy source. This decomposition process enables the production of a wide variety of products, making it increasingly popular in the development of biorefineries. There are three main types of biomass pyrolysis technologies: slow, fast, and flash pyrolysis. Slow pyrolysis involves gradually heating biomass over several hours, resulting in a higher yield of biochar, which is valuable for soil improvement and carbon sequestration, but it produces less bio-oil and syngas compared to other methods. Fast pyrolysis, on the other hand, rapidly heats biomass in under a minute, yielding up to 75% bio-oil, which can be refined into liquid fuels or chemicals, with smaller amounts of syngas and biochar. Flash pyrolysis heats biomass extremely quickly (within seconds) and is designed to maximize bio-oil production, producing very little biochar [28, 29]. Pyrolysis is applicable to a wide variety of biomass feedstocks, including wood, agricultural waste, and organic residues, leading to a high-value production process. Additionally, biochar from pyrolysis holds great potential for use in carbon sequestration [30].

According to previous studies, the pyrolysis process involves numerous complex reactions and various gas products. Gases such as CH_4, C_2H_2, CO, NO, CO_2, C_6H_6, and C_7H_8 are monitored. Analyzing CO is challenging with GC-MS because CO and N_2 share the same molecular weight and mass-to-charge ratio (m/z). To address this, argon is used as the purge gas to create an environment nearly free of N_2. **Figure 3** presents sample data, showing gas species during pyrolysis of a 3/8-inch poplar

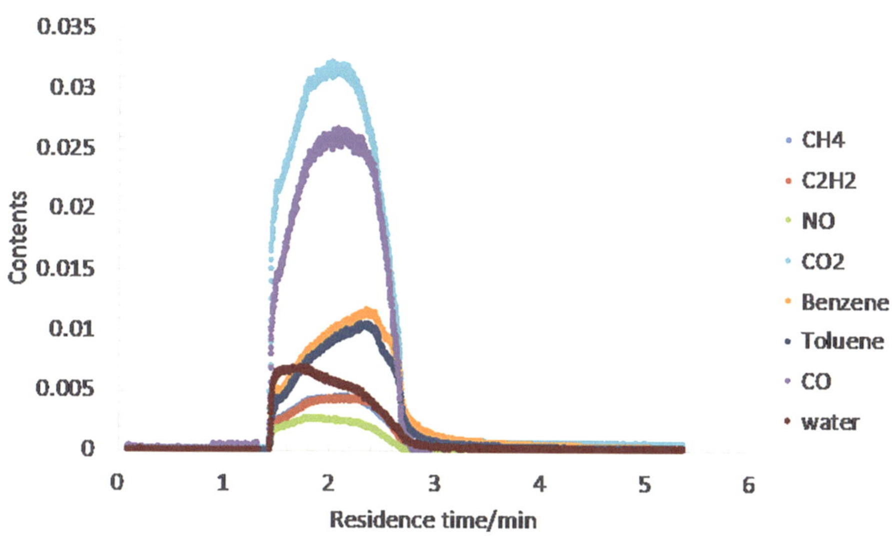

Figure 3.
Compositions and concentrations of pyrolysis gases for 3/8-inch poplar wood in 50% CO_2 and balanced argon at 1150°C [1].

particle in 50% CO_2 and balanced argon at 1150°C. The data indicates that H_2O is released first, followed quickly by benzene and toluene, then C_2H_2, CH_4, and NO. The sharp decline at the end of the process reflects limitations in sampling heavy gases, not a sudden end to pyrolysis. As a matter of fact, pyrolysis gradually slows, with heavier compounds contributing more to mass loss in the form of tar [1].

Biomass pyrolysis is still facing quite several technical challenges. Sizeable energy input, high initial capital cost, process scaling-up limitations, and tar formation issues are associated with biomass pyrolysis. Moreover, the market for pyrolysis byproducts such as biochar, syngas, and bio-oil is still developing, and demand can be limited. This can affect the economic viability of the process, especially if the market for these products does not grow as anticipated [31, 32].

Despite all the technical challenges and issues, biomass pyrolysis is employed in producing biofuels, managing waste, improving soil quality through biochar, and generating renewable energy. It plays a crucial role in advancing sustainable, circular bio-economies by providing alternatives to fossil fuels and lowering carbon emissions.

Thermochemical conversion is a vital component of the renewable energy sector, offering scalable and efficient ways to transform biomass into valuable energy and materials. As technology continues to evolve, this conversion process has the potential to significantly reduce reliance on fossil fuels and contribute to sustainable energy systems.

2.2 Biochemical conversion

Biochemical conversion of biomass involves using biological agents, such as microorganisms and enzymes, to break down organic materials into biofuels or other valuable products. This process typically employs pathways such as anaerobic digestion and fermentation [8, 33, 34]. Both biochemical and thermochemical methods are effective for energy recovery from biomass, as they utilize the entire biomass without needing to isolate specific macromolecules. However, biochemical conversion offers distinct advantages, particularly its ability to handle wet biomass, operate under ambient temperature and pressure, and exhibit greater selectivity for desired products, making it a more sustainable and environmentally friendly option compared to thermochemical processes. Unlike thermochemical conversion, which uses heat, biochemical conversion relies on biological agents. Consequently, the operating condition of biochemical conversion is much more moderate. Common processes of biochemical conversion include anaerobic digestion and fermentation [35]. These processes can convert biomass into various useful products such as bioethanol, biogas, hydrogen, and other chemicals. For instance, anaerobic digestion is commonly used to generate methane-rich biogas, a renewable energy source derived from organic waste [36]. On the other hand, fermentation is primarily applied to convert sugars into bioethanol, a widely used biofuel [37, 38]. Both methods play a critical role in producing sustainable energy from organic materials, helping reduce dependency on fossil fuels and contributing to a more environmentally friendly energy system (**Figure 4**).

2.3 Anaerobic digestion

Anaerobic digestion is a process in which microorganisms break down organic material, such as agricultural waste, food waste, or animal manure, in the absence of

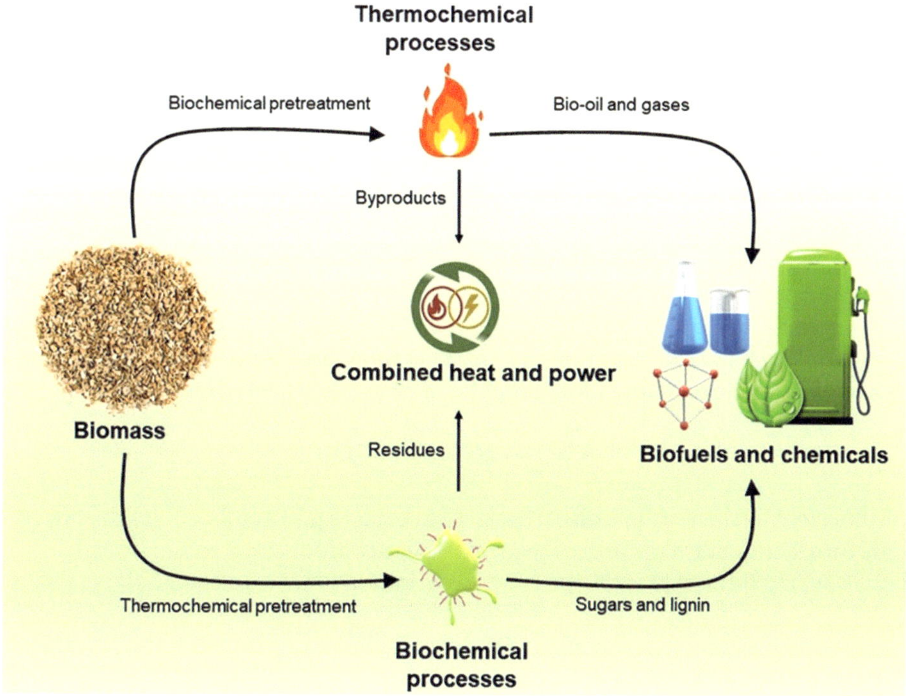

Figure 4.
Scheme of hydrothermal and biochemical routes in biomass utilization from a circular economy perspective [39].

oxygen. The main product of this process is biogas, primarily composed of methane (CH_4) and carbon dioxide (CO_2). This biogas can be used as a renewable energy source for electricity, heating, or even as vehicle fuel [40, 41].

Anaerobic digestion consists of four key stages, catalyzed by different microorganisms, whereas hydrolysis is considered as the rate-determining step [39, 42]:

1. Hydrolysis: Complex organic compounds, such as carbohydrates, fats, and proteins, are broken down into simpler molecules such as sugars, amino acids, and fatty acids. This step is often the slowest and determines the overall rate of digestion [42, 43].

2. Acidogenesis: Fermentative bacteria convert the products of hydrolysis into intermediate compounds, such as volatile fatty acids, alcohols, and other organic molecules [44].

3. Acetogenesis: Homoacetogenic bacteria convert volatile fatty acids and alcohols into acetate, hydrogen, and carbon dioxide [45–47].

4. Methanogenesis: Methanogenic bacteria convert acetate, hydrogen, and carbon dioxide into methane and additional carbon dioxide, to generate biogas [48, 49].

5. Anaerobic digestion is extensively used in waste management and renewable energy production because it reduces waste volume and produces useful byproducts such as biogas and nutrient-rich digestate that can be used as fertilizer.

2.4 Fermentation

Fermentation is a process where microorganisms, such as bacteria or yeast, break down organic materials, typically sugars, into simpler compounds, usually in the absence of oxygen. This process provides energy to the microorganisms and produces byproducts such as alcohol, gases (such as carbon dioxide), or acids, depending on the type of fermentation involved [50, 51]. Fermentation is generally categorized into several types, including alcoholic, lactic acid, and acetic acid fermentation, based on the microorganisms involved, the end products, and the conditions under which the process occurs [39, 52].

Alcoholic fermentation is the technique for yeasts convert sugars (such as glucose) into ethanol (alcohol) and carbon dioxide. This is the process used in brewing beer, making wine, and producing bioethanol from biomass [53, 54]. Reaction is:

$$C_6H_{12}O_6 \rightarrow 2C_2H_5OH + 2CO_2 \tag{1}$$

Lactic acid fermentation utilizes certain bacteria (e.g., Lactobacillus) to convert sugars into lactic acid. This process occurs in yogurt production and in human muscles during intense exercise [55, 56]. Reaction is:

$$C_6H_{12}O_6 \rightarrow 2C_3H_6O_3 \tag{2}$$

Acetic acid fermentation is the process in which acetic acid bacteria (e.g., Acetobacter) convert ethanol into acetic acid (vinegar production) in the existence of oxygen [57]. Reaction is:

$$C_2H_5OH + O_2 \rightarrow C_3COOH + H_2O \tag{3}$$

Fermentation is a key process in the production of bioethanol, which serves as an important renewable fuel. In addition, fermentation technique has played a vital role in food and beverage industry, and pharmaceutical production. This technology provides sustainable and versatile solutions to food production, energy crisis and environmental waste pollution. Despite the importance of the biochemical conversion offers in the production of biofuels, food and pharmaceutical production, renewable energy, waste reduction, and many other fields, there are still a lot of difficulties and challenges, which are preventing it from being commercialized in a large-scale. One major limitation of biochemical conversion is its inability to efficiently process biomass with high hydraulic retention times, as well as the emission of greenhouse gases during the process. Additionally, biomass requires pre-treatment before undergoing biochemical conversion, adding complexity to the system. During fermentation, the biomass must also undergo enzymatic hydrolysis to break down polysaccharides into monosaccharides. Furthermore, a significant gap exists between proven fermentation technologies and their implementation in large-scale biorefineries, with technical and economic barriers, such as selecting the right conversion process and cost considerations, posing obstacles [34, 58–60].

2.5 Physiochemical conversion

Physicochemical conversion of biomass refers to a set of integrated processes that utilize both physical and chemical methods to transform organic materials (such as agricultural residues, wood chips, or food waste) into valuable products such as biofuels, chemicals, or energy. This conversion approach blends mechanical, thermal, and chemical techniques, optimizing the breakdown of biomass into simpler components that can be more efficiently used for energy production or further refined into other materials. The methods are distinct from purely biochemical or thermochemical approaches because they integrate multiple stages of treatment to enhance the breakdown of biomass, optimize the efficiency of subsequent energy extraction processes, and maximize product yields [61–63].

Physicochemical characterization of the biomass properties are essential for this technique. In the physicochemical conversion of biomass, physical methods such as mechanical size reduction and mixing/agitation play a crucial role in preparing the feedstock for subsequent chemical reactions. These physical methods aim to enhance the efficiency of the overall conversion process by increasing the surface area of the biomass and ensuring uniform exposure to chemicals, heat, or biological agents. Mechanical size reduction involves breaking down biomass into smaller, more manageable particles. This can be achieved through techniques such as cutting, grinding, milling, or shredding, depending on the type and hardness of the biomass material. Mixing and agitation are essential steps in physicochemical conversion, particularly in liquid or slurry-based processes. These steps ensure the homogeneous distribution of the biomass particles, solvents, or reactants, which improves the overall reaction efficiency. The physical methods can facilitate handling, decrease the energy cost, and support uniform heat and mass transfer in the chemical reaction process [64].

Chemical techniques, including catalysis, solvent extraction, and super-fluid technology are employed to convert biomass to the desired products. Chemical catalysts are often used to accelerate reactions during physicochemical conversion, such as in biodiesel production where a catalyst is added to oils or fats to enhance the transesterification process. Besides, biomass is treated with solvents to dissolve specific components, such as lignin or cellulose, and extract valuable chemicals. The solvents reduce the energy required for further breakdown of the biomass. In some processes, supercritical fluids (e.g., supercritical CO_2) are used to dissolve and extract materials from biomass. These fluids exhibit both liquid and gas properties under specific conditions, allowing for more efficient separation of biomass components [65, 66].

The major processes applied in physicochemical conversion technologies are:

1. Liquefaction: Biomass is subjected to moderate temperatures (250–350°C) and high pressure in the presence of solvents such as water or alcohols. The goal is to convert solid biomass into a liquid bio-oil, which can be refined into fuels or chemicals. Liquefaction operates at milder conditions than pyrolysis and results in higher liquid yields. This technique can process a wide range of biomass types, including wet feedstocks, and produces bio-oil that is easier to upgrade into fuel than raw biomass [9, 67, 68].

2. Supercritical fluid extraction: This method employs supercritical fluids to extract valuable substances from biomass. For instance, supercritical CO_2 can be used to extract oils from biomass, which can then be processed into biodiesel. Supercritical fluid extraction is highly selective and operates at moderate temperatures, reducing the degradation of sensitive compounds. It is also environmentally friendly because it often uses non-toxic solvents like CO_2 [6, 69, 70].

3. Transesterification: In the production of biodiesel, oils or fats from biomass (such as vegetable oils) are chemically reacted with alcohols (usually methanol or ethanol) in the presence of a catalyst (such as sodium hydroxide). This technique results in biodiesel and glycerol as byproducts. Transesterification is a relatively simple and well-established process, scalable for industrial biodiesel production. It also yields high-quality fuel that can replace conventional diesel in many applications [71, 72].

4. Solvolysis: Biomass is dissolved in solvents, often organic, to break down its complex structure (like lignocellulose) into simpler chemical building blocks. This process helps separate lignin from cellulose and hemicellulose, allowing for the production of biofuels, chemicals, or bio-based materials. This method can be tailored to specific biomass types and the desired end-products and can work well with complex feedstocks [73, 74].

By integrating physical and chemical techniques, these processes can achieve higher yields and greater versatility than many traditional conversion methods, although their economic viability and energy efficiency continue to present challenges for large-scale deployment. However, every coin has two sides. The physicochemical processes, particularly those utilizing supercritical fluids or operating at high-pressure/temperature conditions, require substantial energy input, which can ultimately offset the environmental benefits of biomass conversion. Furthermore, the need for pre-treatment (physical and chemical methods) in most physicochemical processes to enhance the behavior of chemical agents, and increase the overall efficiency is adding complexity to the operational design, and thus leading to a high cost for the process [3, 39, 75].

3. Conclusions

Biomass conversion technologies are a vital component of the global transition to renewable energy and a more sustainable future. By harnessing the energy and material potential of organic matter, these technologies offer pathways to produce biofuels, bioenergy, and bioproducts that can significantly reduce reliance on fossil fuels, mitigate greenhouse gas emissions, and promote a circular economy.

Thermochemical, biochemical, and physicochemical conversion processes each provide unique advantages and challenges, with the choice of technology depending on factors such as feedstock type, desired end-products, and scalability. Thermochemical methods like combustion, gasification, and pyrolysis are highly efficient for large-scale energy production, while biochemical processes such as anaerobic digestion and fermentation are more suited to producing biofuels like biogas and ethanol from biological sources. Physicochemical approaches like transesterification are valuable for converting oils and fats into biodiesel, offering a cleaner alternative to traditional fossil fuels.

Despite their potential, biomass conversion technologies face technical, economic, and environmental challenges that must be addressed to achieve widespread adoption. Feedstock variability, process inefficiencies, and competition with conventional energy sources remain key hurdles. However, continued research and development, along with supportive policies and investments, are paving the way for more efficient, cost-effective, and environmentally sustainable solutions. To overcome the barriers for the development of biomass conversion technology, governments worldwide are implementing a variety of regulations, incentives, and frameworks to promote the use of biomass as a renewable resource. These include subsidies for biomass power plants, tax credits for biofuel producers, and grants for research and development in advanced biomass conversion technologies. Many countries and regions have established renewable energy mandates and targets that deem biomass as a key contributor, which is to push the adoption of biomass-based energy generation and biofuels. All of these policies focus on encouraging investment, ensuring environmental sustainability, and integrating biomass into the broader energy mix to meet climate and energy goals [76–80].

In the future, advancements in biomass conversion technologies hold immense promise. Next-generation processes, integrated biorefineries, and innovative approaches to bio-based products can unlock even greater opportunities for reducing waste, enhancing energy security, and driving economic growth. As these technologies mature, they will play a crucial role in building a cleaner, greener, and more resilient energy system that supports both human well-being and environmental preservation.

Acknowledgment

The authors acknowledges the use of ChatGPT for language polishing of the manuscript.

Conflict of interest

The authors declare no conflict of interest.

Author details

Ruochen Wu
BASF, Huntstville, USA

*Address all correspondence to: wuruochenljty@gmail.com

References

[1] Wu R. Experimental and Modeling of Biomass Char Gasification. Doctoral dissertation, Brigham Young University; 2020

[2] Parikka M. Global biomass fuel resources. Biomass and Bioenergy. 2004;**27**(6):613-620

[3] Ahorsu R, Medina F, Constantí M. Significance and challenges of biomass as a suitable feedstock for bioenergy and biochemical production: A review. Energies. 2018;**11**(12):3366

[4] Mignogna D et al. Biomass energy and biofuels: Perspective, potentials, and challenges in the energy transition. Sustainability. 2024;**16**(16):7036

[5] Vardon DR et al. Realizing "net-zero-carbon" sustainable aviation fuel. Joule. 2022;**6**(1):16-21

[6] Demirbaş A. Biomass resource facilities and biomass conversion processing for fuels and chemicals. Energy Conversion and Management. 2001;**42**(11):1357-1378

[7] Ghoniem AF. Needs, resources and climate change: Clean and efficient conversion technologies. Progress in Energy and Combustion Science. 2011;**37**(1):15-51

[8] Adams P et al. Biomass conversion technologies. In: Greenhouse Gas Balances of Bioenergy Systems. Elsevier, Academic Press; 2018. pp. 107-139

[9] Brown RC. Thermochemical Processing of Biomass: Conversion into Fuels, Chemicals and Power. John Wiley and Sons; 2019

[10] Pandey A et al. Advances in Thermochemical Conversion of Biomass-introduction. In: Recent Advances in Thermochemical Conversion of Biomass. Elsevier; 2015. pp. 3-30

[11] Wang Y, Wu JJ. Thermochemical conversion of biomass: Potential future prospects. Renewable and Sustainable Energy Reviews. 2023;**187**:113754

[12] Chen H, Wang L. Technologies for Biochemical Conversion of Biomass. Academic Press; 2016

[13] Brethauer S, Studer MH. Biochemical conversion processes of lignocellulosic biomass to fuels and chemicals–a review. Chimia. 2015;**69**(10):572-572

[14] Ramos A, Monteiro E, Rouboa A. Biomass pre-treatment techniques for the production of biofuels using thermal conversion methods–a review. Energy Conversion and Management. 2022;**270**:116271

[15] Pant K, Mohanty P. Biomass, conversion routes and products–an overview. Transformation of Biomass: Theory to Practice. 2014:1-30

[16] Wu R, Baxter LL. Dynamic biomass char porosity during gasification: Model compared with data. Energy. 2023;**267**:126603

[17] Wu R, Beutler J, Baxter LL. Biomass char gasification kinetic rates compared to data, including ash effects. Energy. 2023;**266**:126392

[18] Wu R, Beutler J, Baxter LL. Non-catalytic ash effect on char reactivity. Applied Energy. 2020;**260**:114358

[19] Wu R, Beutler J, Baxter LL. Experimental and theoretical biomass char diameter variation during gasification. Energy. 2021;**219**:119431

[20] Wu R et al. Biomass char particle surface area and porosity dynamics during gasification. Fuel. 2020;**264**:116833

[21] Bridgwater T. Biomass for energy. Journal of the Science of Food and Agriculture. 2006;**86**(12):1755-1768

[22] Vassilev SV et al. An overview of the chemical composition of biomass. Fuel. 2010;**89**(5):913-933

[23] McKendry P. Energy production from biomass (part 1): Overview of biomass. Bioresource Technology. 2002;**83**(1):37-46

[24] Field CB, Campbell JE, Lobell DB. Biomass energy: The scale of the potential resource. Trends in Ecology and Evolution. 2008;**23**(2):65-72

[25] Rollinson AN, Karmakar MK. On the reactivity of various biomass species with CO2 using a standardised methodology for fixed-bed gasification. Chemical Engineering Science. 2015;**128**:82-91

[26] Consonni S, Larson E. Biomass-Gasifier/Aeroderivative Gas Turbine Combined Cycles: Part B—Performance Calculations and Economic Assessment. Journal of Engineering for Gas Turbines and Power. 1996;**118**(3):516-525

[27] Liu G et al. Making Fischer–Tropsch fuels and electricity from coal and biomass: Performance and cost analysis. Energy and Fuels. 2011;**25**(1):415-437

[28] Uddin M et al. An overview of recent developments in biomass pyrolysis technologies. Energies. 2018;**11**(11):3115

[29] Al Arni S. Comparison of slow and fast pyrolysis for converting biomass into fuel. Renewable Energy. 2018;**124**:197-201

[30] Yang Q et al. Country-level potential of carbon sequestration and environmental benefits by utilizing crop residues for biochar implementation. Applied Energy. 2021;**282**:116275

[31] Rasaq WA et al. Navigating pyrolysis implementation—A tutorial review on consideration factors and thermochemical operating methods for biomass conversion. Materials. 2024;**17**(3):725

[32] Laird DA et al. Review of the pyrolysis platform for coproducing bio-oil and biochar. Biofuels, Bioproducts and Biorefining. 2009;**3**(5):547-562

[33] Kumar R, Singh S, Singh OV. Bioconversion of lignocellulosic biomass: Biochemical and molecular perspectives. Journal of Industrial Microbiology and Biotechnology. 2008;**35**(5):377-391

[34] Manikandan S et al. Critical review of biochemical pathways to transformation of waste and biomass into bioenergy. Bioresource Technology. 2023;**372**:128679

[35] Awasthi MK et al. A comprehensive review on thermochemical, and biochemical conversion methods of lignocellulosic biomass into valuable end product. Fuel. 2023;**342**:127790

[36] Bharathiraja B et al. Biogas production–a review on composition, fuel properties, feed stock and principles of anaerobic digestion. Renewable and Sustainable Energy Reviews. 2018;**90**(April):570-582

[37] Rastogi M, Shrivastava S. Recent advances in second generation bioethanol production: An insight to pretreatment, saccharification and fermentation processes. Renewable and Sustainable Energy Reviews. 2017;**80**:330-340

[38] Azhar SHM et al. Yeasts in sustainable bioethanol production: A review. Biochemistry and Biophysics Reports. 2017;**10**:52-61

[39] Osman AI et al. Conversion of biomass to biofuels and life cycle assessment: A review. Environmental Chemistry Letters. 2021;**19**:4075-4118

[40] Adekunle KF, Okolie JA. A review of biochemical process of anaerobic digestion. Advances in Bioscience and Biotechnology. 2015;**6**(03):205-212

[41] Uddin MM, Wright MM. Anaerobic digestion fundamentals, challenges, and technological advances. Physical Sciences Reviews. 2023;**8**(9):2819-2837

[42] Patel V, Pandit S, Chandrasekhar K. Basics of Methanogenesis in Anaerobic Digester. Microbial Applications Vol. 2: Biomedicine, Agriculture and Industry. Springer; 2017. pp. 291-314

[43] Menzel T, Neubauer P, Junne S. Role of microbial hydrolysis in anaerobic digestion. Energies. 2020;**13**(21):5555

[44] Kim M et al. Hydrolysis and acidogenesis of particulate organic material in mesophilic and thermophilic anaerobic digestion. Environmental Technology. 2003;**24**(9):1183-1190

[45] Ziemiński K, Frąc M. Methane fermentation process as anaerobic digestion of biomass: Transformations, stages and microorganisms. African Journal of Biotechnology. 2012;**11**(18):4127-4139

[46] Bajpai P, Bajpai P. Basics of anaerobic digestion process. Anaerobic Technology in Pulp and Paper Industry. 2017:7-12

[47] Pan X et al. Deep insights into the network of acetate metabolism in anaerobic digestion: Focusing on syntrophic acetate oxidation and homoacetogenesis. Water Research. 2021;**190**:116774

[48] Krakat N et al. Anaerobic digestion of renewable biomass: Thermophilic temperature governs methanogen population dynamics. Applied and Environmental Microbiology. 2010;**76**(6):1842-1850

[49] Demirel B, Scherer P. The roles of acetotrophic and hydrogenotrophic methanogens during anaerobic conversion of biomass to methane: A review. Reviews in Environmental Science and Bio/Technology. 2008;**7**:173-190

[50] Maicas S. The Role of Yeasts in Fermentation Processes. Microorganisms. 2020;**8**(8):1142

[51] Sarris D, Papanikolaou S. Biotechnological production of ethanol: Biochemistry, processes and technologies. Engineering in Life Sciences. 2016;**16**(4):307-329

[52] De Vuyst L, Leroy F. Functional role of yeasts, lactic acid bacteria and acetic acid bacteria in cocoa fermentation processes. FEMS Microbiology Reviews. 2020;**44**(4):432-453

[53] Van Maris AJ et al. Alcoholic fermentation of carbon sources in biomass hydrolysates by Saccharomyces cerevisiae: Current status. Antonie Van Leeuwenhoek. 2006;**90**:391-418

[54] Lin Y, Tanaka S. Ethanol fermentation from biomass resources: Current state and prospects. Applied Microbiology and Biotechnology. 2006;**69**:627-642

[55] John RP, Nampoothiri KM, Pandey A. Fermentative production of lactic acid from biomass: An overview on process developments and future perspectives. Applied Microbiology and Biotechnology. 2007;**74**:524-534

[56] Li Y et al. Sustainable lactic acid production from lignocellulosic biomass. ACS Sustainable Chemistry and Engineering. 2021;**9**(3):1341-1351

[57] Ross MK. Production of Acetic Acid from Waste Biomass. Texas A&M University; 1998

[58] Appel AM et al. Frontiers, opportunities, and challenges in biochemical and chemical catalysis of CO2 fixation. Chemical Reviews. 2013;**113**(8):6621-6658

[59] Menon V, Rao M. Trends in bioconversion of lignocellulose: Biofuels, platform chemicals and biorefinery concept. Progress in Energy and Combustion Science. 2012;**38**(4):522-550

[60] Siddiki SYA et al. Microalgae biomass as a sustainable source for biofuel, biochemical and biobased value-added products: An integrated biorefinery concept. Fuel. 2022;**307**:121782

[61] Cai J et al. Review of physicochemical properties and analytical characterization of lignocellulosic biomass. Renewable and Sustainable Energy Reviews. 2017;**76**:309-322

[62] Shirkavand E et al. Combination of fungal and physicochemical processes for lignocellulosic biomass pretreatment–a review. Renewable and Sustainable Energy Reviews. 2016;**54**:217-234

[63] Baskar C, Baskar S, Dhillon RS. Biomass Conversion: The Interface of Biotechnology, Chemistry and Materials Science. Springer Science and Business Media; 2012

[64] Manan M, Webb C. Design aspects of solid state fermentation as applied to microbial bioprocessing. Journal of Applied Biotechnology and Bioengineering. 2017;**4**(1):511-532

[65] Singh D et al. A review on feedstocks, production processes, and yield for different generations of biodiesel. Fuel. 2020;**262**:116553

[66] Ambat I, Srivastava V, Sillanpää M. Recent advancement in biodiesel production methodologies using various feedstock: A review. Renewable and Sustainable Energy Reviews. 2018;**90**:356-369

[67] Chornet E, Overend RP. Biomass liquefaction: An overview. Fundamentals of Thermochemical Biomass Conversion. 1985:967-1002

[68] Zhang S et al. Liquefaction of biomass and upgrading of bio-oil: A review. Molecules. 2019;**24**(12):2250

[69] Escobar ELN et al. Supercritical fluids: A promising technique for biomass pretreatment and fractionation. Frontiers in Bioengineering and Biotechnology. 2020;**8**:252

[70] Akalın MK, Tekin K, Karagöz S. Supercritical fluid extraction of biofuels from biomass. Environmental Chemistry Letters. 2017;**15**:29-41

[71] Bhatia SK et al. An overview on advancements in biobased transesterification methods for biodiesel production: Oil resources, extraction, biocatalysts, and process intensification technologies. Fuel. 2021;**285**:119117

[72] Jha S et al. A review of thermochemical conversion of waste biomass to biofuels. Energies. 2022;**15**(17):6352

[73] Duan D et al. Comparative study on various alcohols solvolysis of organosolv lignin using microwave energy: Physicochemical and morphological

properties. Chemical Engineering and Processing-Process Intensification. 2018;**126**:38-44

[74] Basak B et al. Advances in physicochemical pretreatment strategies for lignocellulose biomass and their effectiveness in bioconversion for biofuel production. Bioresource Technology. 2023;**369**:128413

[75] Gallezot P. Catalytic conversion of biomass: Challenges and issues. ChemSusChem: Chemistry and Sustainability Energy and Materials. 2008;**1**(8-9):734-737

[76] Capodaglio AG, Callegari A, Lopez MV. European framework for the diffusion of biogas uses: Emerging technologies, acceptance, incentive strategies, and institutional-regulatory support. Sustainability. 2016;**8**(4):298

[77] Banja M et al. Biomass for energy in the EU–the support framework. Energy Policy. 2019;**131**:215-228

[78] Yana S, Nizar M, Mulyati D. Biomass waste as a renewable energy in developing bio-based economies in Indonesia: A review. Renewable and Sustainable Energy Reviews. 2022;**160**:112268

[79] Stupak I et al. Sustainable utilisation of forest biomass for energy—Possibilities and problems: Policy, legislation, certification, and recommendations and guidelines in the Nordic, Baltic, and other European countries. Biomass and Bioenergy. 2007;**31**(10):666-684

[80] Abdul Malek A, Hasanuzzaman M, Rahim NA. Prospects, progress, challenges and policies for clean power generation from biomass resources. Clean Technologies and Environmental Policy. 2020;**22**:1229-1253

Chapter 2

Biomass to Energy: Harnessing Nature's Potential

Mohammad Ali Sabbaghi

Abstract

This chapter explores a wide range of valuable products that can be derived from biomass. Biomass is an abundant and renewable energy source that can contribute to alleviating the energy crisis and addressing environmental challenges posed by fossil fuels. This energy can be transformed into fuels, chemicals, and other valuable products, serving as a sustainable alternative to fossil fuels. The chapter begins with a discussion on biofuels, such as bioethanol, biodiesel, and biogas, and how they can be produced from various biomass feedstocks, including agricultural residues, forest products, and energy crops. The topic of converting waste into energy will also be examined, along with methods for energy extraction from biomass. Generally, thermochemical processes, particularly gasification, exhibit faster conversion rates and higher efficiencies. When compared to biochemical methods, thermochemical gasification appears to be a more favorable option for extracting energy in the form of synthesis gas. This process not only accelerates the transformation of feedstocks into valuable energy but also enhances the overall yield, making it an attractive alternative for energy production. Biomass gasification produces a combustible gas mixture. The composition and energy content of this gas vary based on factors like the type of gasifying agent used in the process.

Keywords: biomass, energy, pyrolysis, gasification, sustainable source

1. Introduction

Energy holds significant importance in the development of industry and facilitates international trade. For this reason, energy security is of great importance for the advancement and expansion of technology. Producing energy in a safe manner is one of the priorities of human life. The production, consumption, and preservation of energy are essential for achieving sustainable development. Given the negative environmental impacts caused by the consumption of fossil fuels, which release NO_x, CO_2, and SO_x gases in the transportation, industrial, power generation, and residential sectors, there is a growing need for renewable energy sources.

With the increase in population, the dependence on energy and its consumption also rises. The use of fossil fuels can contribute to the rise in global temperatures and cause serious damage to the environment. Additionally, burning these fuels for energy and heat production leads to an increase in greenhouse gases [1].

Biomass is a plentiful and renewable energy source that can help alleviate the energy crisis and the environmental challenges arising from fossil fuels. This type of energy can be converted into fuels, chemicals, and other valuable products, serving as a sustainable alternative to fossil fuels. Biomass refers to biological materials or waste derived from living organisms, which stores solar energy. This energy source typically consists of organic materials that include carbon, hydrogen, and oxygen, and encompasses plant residues, municipal waste, industrial waste, animal waste, and household garbage.

Biomass, in its dry form, typically consists of a carbon content ranging from 30% to 60% and a hydrogen content of approximately 5% to 6%. Other mineral elements such as nitrogen, chlorine, and sulfur contribute to less than 1% of the total biomass composition. This diverse makeup highlights the potential of biomass as a versatile energy resource, capable of reducing dependency on fossil fuels while addressing environmental concerns [2].

The main factors that are important in the utilization of biomass for energy production include heating value, moisture content, and ash content [3].

- Heating value indicates the amount of energy released when biomass is burned.
- Moisture content denotes the amount of water present in the feedstock, such as wood.
- Ash content reflects the percentage of mineral matter present in the biomass.

1.1 The importance of energy production from biomass

Biomass, due to its renewable and sustainable characteristics, can be recognized as an alternative energy source. It is predicted that this resource has the potential to meet 25% of the global energy demand. Based on a report by the U.S. Department of Energy, renewable energies account for about 10% of the total energy consumed in the country, and half of this amount, approximately 50%, is derived from biomass. This showcases the significant role that biomass could play in the transition toward more sustainable energy systems. As the global community is striving to decrease its dependence on fossil energy sources and mitigate climate change, the utilization of biomass presents a viable and effective solution for meeting energy needs while promoting environmental sustainability.

Biomass plays an important role in the energy industry and is recognized as the largest primary energy source in the world after fossil fuels. To produce and utilize biomass effectively, it is essential to have sufficient knowledge about its various types. This information helps us harness biomass for generating electricity, energy, transportation fuels, heating, and other products. Understanding the diverse forms and applications of biomass can lead to more sustainable energy practices and innovations in various sectors, ultimately contributing to a greener future [3].

2. Diverse forms of biomass energy

Biofuels play a crucial role in the transition to a sustainable energy system. Among these fuels, bioethanol and biodiesel have gained attention due to their capacity to lower the emissions of greenhouse gases and enhance energy security.

Bioethanol is primarily produced through the fermentation of carbohydrates found in biomass. These carbohydrates are typically sourced from agricultural residues, food products such as corn and sugarcane, and energy-rich resources. During the fermentation process, microorganisms convert sugars into ethanol. This bioethanol can then be blended with gasoline and utilized in vehicles, resulting in a decreased reliance on petroleum and a reduction in carbon emissions.

Biodiesel is created using a method known as transesterification. This technique involves the conversion of triglycerides present in plant oils or animal fats are converted into fatty acid methyl esters. This renewable fuel can be utilized as is or mixed with traditional diesel, making it an attractive option for those looking to reduce their carbon footprint.

The essential ingredients required for the creation of biodiesel include soybean oil, canola oil, and waste cooking oils. Since biodiesel can be produced economically, it presents opportunities for rural development. Farmers can increase their income by cultivating oilseeds or collecting waste oils.

In addition to bioethanol and biodiesel, biogas plays a significant role as a new biomass fuel in the future of renewable energy. Biogas is generated by the anaerobic breakdown of organic materials such as manure, food waste, and agricultural residues. During this natural process, microorganisms break down organic matter, resulting in a mixture primarily composed of methane and carbon dioxide. This biogas can be utilized directly for producing heat and generating electricity, or converted into biomethane for injection into natural gas networks. The use of biogas not only acts as a sustainable energy source but also contributes to waste reduction and nutrient recovery, exemplifying the principles of a circular economy.

The diversity of raw materials in biomass is a key factor in the production of biofuels and its potential to provide sustainable energy solutions. Agricultural residues, such as wheat straw and rice husks, offer substantial sources of materials that are often discarded, contributing to improved resource efficiency. Forest products, including wood chips and sawdust, enable the utilization of waste from the wood industry. Additionally, crops that are specifically cultivated for high-yield biomass create favorable conditions for the development of future biofuels [4].

2.1 Biomass extraction considerations

Regarding biomass as a renewable resource, it is crucial to consider the environmental and socioeconomic impacts of its production. To ensure biomass is provided sustainably, it is essential to guarantee that land use does not interfere with food production and that negative environmental effects are minimized. These considerations are significant for strengthening the biofuel industry and supporting global sustainability objectives.

Policy-making, research, and public awareness-raising play crucial roles in enhancing the sustainability and acceptance of biofuels. These efforts help ensure that the benefits of these fuels are realized in a way that simultaneously promotes economic growth and protects the environment. By developing effective policies and investing in research, stakeholders can better understand the potential of biofuels and address any challenges associated with their production and use. Additionally, increasing public awareness about the advantages and importance of biofuels is vital for fostering community support and encouraging sustainable practices. Overall, a collaborative approach that emphasizes these key roles will be instrumental in advancing the biofuel sector and achieving a balanced relationship between economic development and environmental stewardship [5].

Central to the process of converting biomass is the generation of biochemicals, which play a vital role in today's economy. Chemicals derived from biomass function as essential components for numerous industrial uses. These substances can be synthesized using biotechnological methods such as fermentation or through thermochemical methods like gasification. The innovative potential of microorganisms in these biotechnological processes facilitates the eco-friendly production of valuable platform chemicals from basic biomass resources. For example, cellulose can be broken down into sugars, which are then fermented to produce bioethanol, an attractive substitute for conventional gasoline. Moreover, specialty chemicals, which find applications across various sectors including pharmaceuticals and beauty products, can also be derived from biomass. The production of fine chemicals, characterized by their high purity and targeted functionalities, is increasingly taking place through biocatalysts—an advancement that promotes more sustainable manufacturing practices. By harnessing biomass for chemical production, we can reduce dependency on fossil fuels while fostering the development of greener methodologies, thereby addressing significant environmental issues.

An increasing issue facing global industries is the rising amount of plastic waste and the negative effects of synthetic materials on the environment. A potential answer lies in the use of polymers and composites derived from biomass, which present renewable options that can supplant traditional fossil fuel-based materials. One notable example is polylactic acid (PLA), sourced from corn starch, which has become widely used in packaging as an eco-friendly alternative to standard plastics. Likewise, lignin, extracted from wood and agricultural by-products, can be transformed into durable, biodegradable composites suitable for various applications in construction and textiles.

Exploring the topic of biomass requires a thorough examination of the innovative waste-to-energy technologies that play a crucial role in its transformation. Different biomass waste sources encompass agricultural leftovers, forestry residues, and municipal waste, present ample opportunities for energy recuperation. Techniques such as anaerobic digestion and pyrolysis enable the conversion of these waste materials into useful energy outputs such as heat, electricity, and fuels for transportation. Anaerobic digestion is a biochemical process in which organic materials are decomposed in the absence of oxygen, yielding biogas as a result. This biogas, predominantly made up of methane, can serve multiple purposes, including heating, providing combined heat and power, or even being refined into biomethane for use in vehicles. Conversely, pyrolysis encompasses the thermal breakdown of biomass at high temperatures, resulting in products such as bio-oil, syngas, and biochar, thereby expanding the array of available energy resources. By effectively managing biomass waste, we can reduce the burden on landfills while promoting a more sustainable and diverse energy future [6].

3. General strategies for biomass processing

Extensive research has been conducted on methods of energy extraction from biomass. This resource is utilized for the production of fuels and chemicals through two main technologies: biochemical and thermochemical methods. Biochemical conversion includes processes such as digestion for biogas production and fermentation for ethanol generation. However, there are concerns regarding the economic viability of these methods. The increased production of fuels from crops such as wheat, barley, and corn can lead to a rise in food prices, sparking debates about the competition between food and fuel [7].

In the field of enzymatic hydrolysis as a biochemical method, several challenges exist. These challenges include the low activity of commercial enzymes, the high costs associated with these enzymes, and an insufficient understanding of the biochemistry, structure, and function of the enzymes. Addressing these challenges is crucial for making enzymatic hydrolysis a more viable option for biomass conversion.

Thermochemical methods based on biomass are recognized as an alternative option for the production of biofuels and chemical materials. These processes can be utilized to generate electricity as well as biofuels. The main methods in thermochemistry include direct combustion, pyrolysis (thermal decomposition), liquefaction, gasification, and carbonization. Overall, these thermochemical pathways demonstrate higher efficiency and yield compared to biochemical methods.

Pyrolysis refers to the process of producing liquid at medium temperatures and with short residence times in an oxygen-free environment. The pressure during this process is typically equal to atmospheric pressure, and combustion does not occur due to the lack of oxygen. Instead, the chemical compounds found in organic material, such as cellulose, hemicellulose, and lignin, are converted into flammable gases. These gases can then be transformed into combustible liquids, resulting in products such as pyrolysis oil or bio-oil. However, to achieve high heating rates, this process requires the drying of the materials [8].

As an alternative, hydrothermal liquefaction (HTL) utilizes wet biomass and eliminates the need for drying, which often entails high energy consumption.

Thermal gasification refers to the production of gaseous products in the presence of factors such as air, oxygen, and steam. This process is based on the partial oxidation of biomass and results in the generation of syngas, which includes hydrogen, carbon dioxide, methane, and carbon monoxide. Gasification serves as a crucial method for converting biomass into valuable energy carriers, as it allows for the efficient utilization of organic materials. By harnessing the chemical energy present in biomass, thermal gasification facilitates the transition toward more sustainable

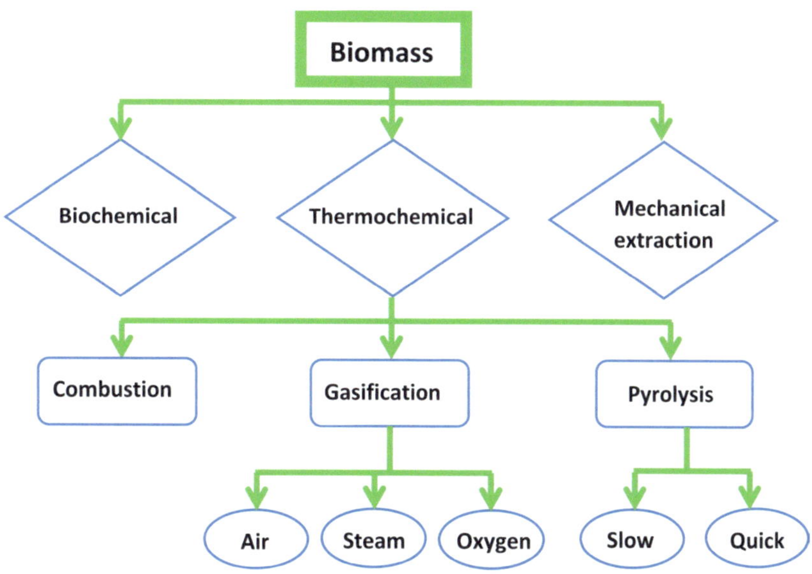

Figure 1.
Various strategies of energy extraction from biomass.

energy systems. Furthermore, the syngas produced can be utilized in various applications, including electricity production, chemical production, and as a fuel for transportation [9].

Biomass with high moisture content requires drying to ensure effective energy conversion during the thermal gasification process. However, in supercritical water gasification (SCWG), there is no need for drying, and wet biomass can be utilized directly. Generally, thermochemical processes, particularly gasification, exhibit faster conversion rates and higher yields. In comparison with biochemical methods, thermochemical gasification appears to be a better option for extracting energy in the form of synthesis gas [10].

Various types of energy extraction strategies from biomass are shown in **Figure 1**.

4. Gasification

Gasification stands out as a highly efficient approach for transforming biomass and waste into energy within the realm of thermochemical processes. In this process, biomass is combusted in the presence of air, and the chemical energy contained within it is transformed into heat, electricity, or mechanical power. However, one of the significant challenges associated with this method is the high emission of pollutants and the generation of large amounts of ash [11].

The gasification of biomass is an endothermic process, which means it requires heat to proceed. Depending on the source of heat, this process can be categorized as either autothermal or non-thermal. In autothermal gasification, heat is produced through the partial oxidation of biomass occurring within the gasifier itself. On the other hand, in non-thermal gasification, heat is supplied from an external source, which often presents challenges related to the transfer of heat to the area where the conversion to high-temperature gas takes place. This distinction between the two types of gasification processes is crucial for understanding their operational dynamics and efficiency in producing gas from biomass feedstock [8, 12, 13].

The primary output of biomass gasification consists of a blend of flammable substances, comprising carbon monoxide, methane, nitrogen, hydrogen, and lighter hydrocarbons, along with small amounts of heavier hydrocarbons. This gas blend is known as producer gas, and its characteristics and energy content depend on various factors, including the type of agents employed in the gasification process. Following a purification and refining procedure aimed at eliminating or decreasing contaminants to minimal concentrations, the produced synthesis gas is mainly composed of hydrogen and carbon monoxide. This gas can be harnessed as a fuel source in internal combustion engines or gas turbines for generating heat and electricity. Moreover, synthesis gas acts as a key starting material for creating a range of chemical substances and biofuels [14–16].

The main advantage of biomass gasification over other thermochemical conversion technologies is the utilization of very high temperatures. These elevated temperatures lead to the production of higher quality and purer syngas as unwanted impurities are significantly removed in the process. Additionally, biomass gasification demonstrates a higher efficiency compared to other technologies.

Research conducted by Gírio et al. [17] indicates that several technologies for converting biomass waste into energy and advanced biofuels have been examined. These studies rely on multiple performance metrics, with efficiency being a key focus. The investigators have found that gasification exhibits comparable efficiency to combustion and surpasses several biomass conversion methods, including anaerobic digestion, fermentation, and pyrolysis.

In addition, regulatory measures set by government agencies and global institutions enhance the significance of gasification. This method allows for the utilization of diverse raw materials, such as lignocellulosic substances, plastics, and other waste products, regardless of their moisture levels, tailored to the specific gasification technique used [18].

Consequently, gasification processes exhibit significant flexibility concerning the raw materials used. Unlike other biomass conversion technologies, such as bioethanol production *via* fermentation, which require land-use changes, gasification does not necessitate such transformations. This characteristic makes gasification a more sustainable option, as it can utilize waste materials that might otherwise contribute to environmental pollution without competing with food production or causing deforestation.

4.1 Syngas

Converting biomass into gas comes with limitations, including logistical challenges. Because biomass has a lower density in comparison with fossil fuels, transporting it becomes more difficult. Furthermore, gasification typically results in the production of syngas with a diverse combination of gaseous components and unwanted impurities. This increases the need for cleaning and purification stages, which in turn adds complexity and raises costs.

The formation of tar is also a major concern in biomass gasification. Tar accumulation can obstruct machinery, leading to diminished efficiency in processes and a decrease in the quality of syngas, which in turn elevates operational expenses. Furthermore, the emission of tar into the environment may present health hazards and ecological concerns, given that numerous compounds within tar are known to be carcinogenic [19].

One approach to minimize tar production is to operate at elevated temperatures or utilize catalysts. Recently, advancements have been made to address this challenge, particularly through the application of microwave technology [20, 21]. A major issue in the gasification process is the high concentration of impurities found in syngas derived from biomass. These impurities encompass a wide range of substances, including various tar compounds, particulates, nitrogen-containing compounds such as ammonia and hydrogen cyanide, as well as sulfur-bearing inorganic substances such as hydrogen sulfide, carbon disulfide, and carbonyl sulfide. Additionally, halides and halogens such as hydrogen chloride and chlorine, together with alkali metals such as sodium and potassium, contribute to the contamination. These pollutants are generated during the thermochemical conversion of biomass and can appear in varying concentrations within the resulting syngas. The presence of these contaminants can lead to numerous technical and operational issues, including equipment degradation caused by hydrogen sulfide, fouling resulting from tars, and deactivation of catalysts due to substances such as hydrogen sulfide, ammonia, hydrogen chloride, and trace metals [22, 23].

There are numerous approaches available for improving syngas quality by minimizing contaminant levels, which can be classified into primary and secondary methods. Primary methods focus on the immediate reduction of contaminants, such as tars, right within the gasification process and the reactor itself. In contrast, secondary methods address the treatment of syngas after its formation [23]. *In situ* techniques rely on careful management of operational parameters and the incorporation of catalysts or additives during the gasification process to limit the production of contaminants like tar. In comparison, secondary processes do not disrupt the gasification reaction itself; rather, they aim to remove impurities from the syngas downstream through various physical and chemical means. Despite the different methods available, relying solely on a single technique for effective contaminant removal remains a difficult task in practice. A more successful strategy often involves the combined use of both *in situ* and post-gasification treatment methods [24].

The construction of a gasifier is influenced by multiple elements, such as the fuel being used, its dimensions and form, the moisture levels and ash presence in the input material, as well as the specific uses for the syngas produced. To accommodate these needs, a range of gasifier types has been created. These encompass fixed bed reactors (available in updraft, downdraft, and crossflow designs), fluidized bed reactors (which include bubbling, circulating, two-stage, and multi-stage options), bubbling flow reactors (featuring top and side feeding methods), and plasma reactors. In the context of business uses, nearly 75% of gasifiers are of the downdraft type, while 20% utilize fluidized bed technologies, and the remaining 5% are allocated to other designs. **Figure 1** schematically illustrates the different types of gasifiers [25, 26]. Different types of gasifiers are shown in **Figure 2**.

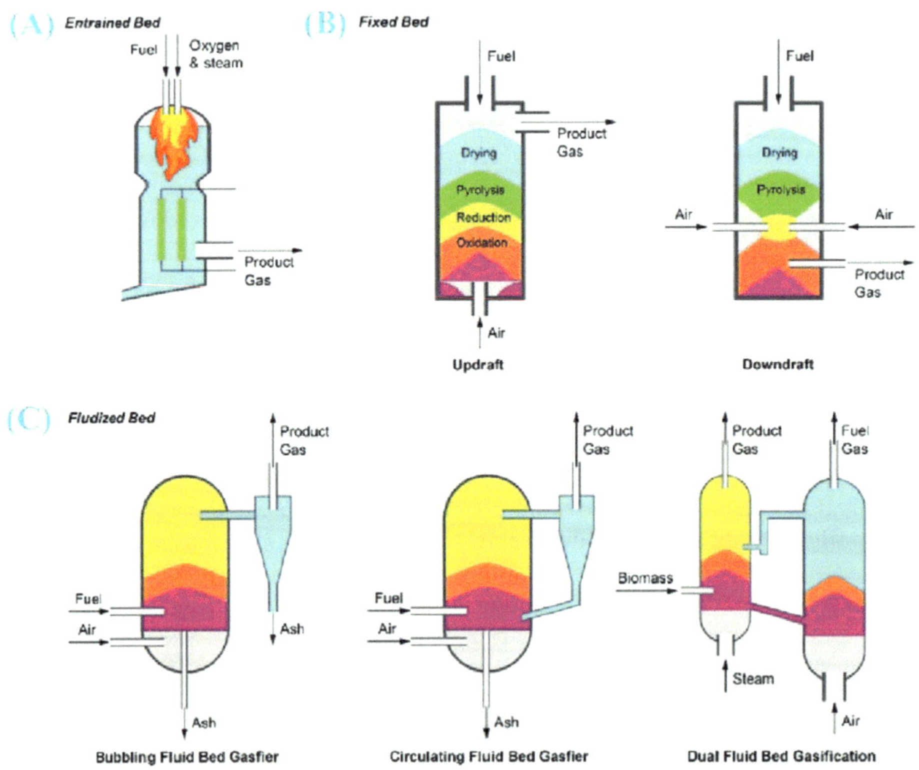

Figure 2.
Various types of gasifiers.

4.2 Synthesis gas cleaning processes

There are two primary classifications based on the kind of working fluid involved: purification of wet gases and purification of dry gases. Wet gas purification involves the use of a liquid medium, often water or other solvents, to absorb or react with pollutants, thereby removing them from the gas stream. This method is especially effective for soluble contaminants and can increase the overall efficiency of the purification process. On the other hand, dry gas purification employs solid adsorbents or other dry media to capture particulates and gaseous pollutants without the need for a liquid phase. Each of these approaches has its own advantages and limitations, making the choice of purification method dependent on the specific composition of the synthesis gas and the nature of the contaminants present.

4.2.1 Wet gas cleaning (WGC)

WGC is recognized as a highly effective and economical technology for purifying gases, successfully minimizing the levels of specific pollutants to acceptable thresholds. This method is particularly favored for the elimination of impurities from raw syngas, owing to its proven dependability and superior performance. Typically, WGC operates at temperatures under 100°C [27, 28]. However, it faces challenges, as the gasification process is carried out at higher temperatures ranging from 700 to 900°C, leading to a decrease in energy efficiency due to the required cooling of the syngas. Furthermore, there are additional expenses linked to the management or disposal of the pollutants collected during the cleaning operation [27].

4.2.2 Dry gas cleaning (DSC)

Recently, there has been an increasing emphasis on techniques for DGC owing to its notable advantages concerning energy efficiency and environmental impact. This technique is particularly impressive for its ability to clean and condition gasification by-products at temperatures that are similar to those in the gasification stage, which helps to minimize energy loss overall. Furthermore, DGC presents additional benefits, such as reducing waste outputs and the possibility of transforming certain pollutants into useful by-products. For example, tar can be converted into useful combustible gases such as carbon monoxide (CO), methane (CH_4), and hydrogen (H_2). DGC can be divided into three primary categories: mechanical-physical methods aimed at particle removal, thermal methods focused on eliminating tar, and catalytic methods designed to remove a range of contaminants, including tar, sulfur, alkali, nitrogen, and chlorine [29].

4.3 Advances in gasification method

Recent advancements in biomass gasification have included significant progress in integrated gasification methods, supercritical water gasification, and cyclic chemical gasification. These methods show significant potential due to their objective of reducing the costs of syngas generation and improving the efficiency of gasification systems. This goal is achieved by the combination of various processes, highlighting the potential for fundamental changes in this field.

4.4 Carbon capture technology

With the transition toward a net-zero energy framework, carbon capture unit (CCU) technologies are becoming increasingly important. Utilizing CCU technologies in biomass gasification facilities is essential for minimizing carbon emissions, which helps transform gasification methods into more sustainable alternatives. Additionally, Carbon Capture Storage and Utilization (CCSU) technologies enable the efficient extraction and conversion of carbon dioxide into valuable renewable products such as methanol, methane, and liquid fuels. This advancement is pivotal in the effort to decarbonize sectors with significant emissions, including transportation, cooling, and heating, as well as the chemical industry [30].

The transformation of syngas into renewable fuels and chemicals encompasses processes such as methanation, methanol production, and Fischer-Tropsch synthesis. While there are certain scenarios where a minor concentration of CO_2 (below 10% by volume on a dry basis) can be advantageous, most syngas conversion methods require the elimination of CO_2 to enhance product quality, optimize reaction rates, and avert the poisoning of catalysts. Additionally, removing CO_2 from syngas is often necessary to reduce its overall volume, which allows for the utilization of more compact processing equipment.

Modifying the H_2-to-CO ratio in syngas is essential, which can be accomplished through water-gas shift (WGS) reactions or by introducing additional hydrogen [31]. In gasification systems, CO_2 capture tends to be more straightforward and cost-effective relative to other industries, primarily due to the elevated pressure of the resulting syngas, resulting in a higher concentration of CO_2. The captured carbon dioxide can be utilized on-site for the production of valuable products, as previously mentioned, or it can be marketed for additional financial benefits. Potential uses for CO_2 include its incorporation into building materials, its role in carbonated drinks, the mineralization of hard water, its function in flame retardants such as fire extinguishers, and its application as a pressurized gas in pneumatic tools and machinery.

The combination of gasification with the combined cycle (CC) can be beneficial both environmentally and economically. However, additional investigation is required to completely grasp and illustrate these mechanisms. Recently, several studies on this topic have been published, contributing to our understanding of the potential advantages and operational efficiencies of integrating gasification technology with combined cycle power generation. These studies could pave the way for more sustainable energy solutions, highlighting not only the environmental benefits but also the potential for cost savings in energy production. The ongoing research efforts aim to bridge the knowledge gaps and optimize the performance of this integrated approach [32, 33].

5. Biomass gasification thermodynamic equations

The interaction of biomass with steam generates a variety of gases, including hydrogen, carbon monoxide, carbon dioxide, methane, and additional water vapor. In the gasification of biomass, key reactions take place, encompassing the transformation of biomass, the water-gas reaction, the water-gas shift, and the generation of methane. A gasifier reactor operating under isothermal conditions efficiently transforms the feedstock into synthesis gas [34].

The biomass gasification process is carried out using water vapor as a reactant. This gasification reaction is defined as shown in Eq. 1. Biomass gasification involves the thermal decomposition of organic materials in the presence of limited oxygen and steam, resulting in the production of syngas, mainly composed of hydrogen, carbon monoxide, and small amounts of other gases. This process not only helps in converting biomass into a more usable form of energy, but also plays a significant role in reducing waste and promoting renewable energy sources. Understanding the dynamics of the gasification reaction is crucial for optimizing the efficiency of biomass-to-energy conversion processes [34].

$$CH_xO_y + zH_2O \rightarrow n_{H2}H_2 + n_{co}CO + n_{CO2}CO_2 + n_{H2O}H_2O + n_{CH4}CH_4 \quad (1)$$

The formulas pertain to the molar proportions of hydrogen relative to carbon and oxygen relative to carbon found in biomass. Moreover, the variable z denotes the overall moisture content in the biomass, in conjunction with the added water vapor, while n_i signifies the production of synthetic gas from the biomass gasification process. Stoichiometric total biomass ratio (STBR) is identified as one of the key performance indicators for biomass gasification, as indicated in Eq. 2.

$$STBR = \frac{\dot{m}_{moisture} + \dot{m}_{steam,in}}{\dot{m}_{dry\,biomass}} \quad (2)$$

The efficiency of cold gas is defined according to Eq. 3, which relates the lower heating value of the gases exiting the gasifier reactor to the total lower heating value of the biomass and the heat input to the reactor. This relationship allows us to quantify how effectively the gasifier converts biomass into usable gas, taking into consideration both the energy content of the available biomass and the thermal energy provided to the system. Understanding this efficiency metric is important for optimizing gasification processes [35].

$$\eta_c = \frac{LHV_{flue\ gas}}{LHV_{biomass} + Q_{in}} \times 100 \quad (3)$$

The exergy destruction rate within the biomass gasifier's control volume is articulated in Eq. 4. In this equation, ($\dot{Ex}_{steam}$) and ($\dot{Ex}_{syngas}$) account for the complete physical and chemical exergy associated with the incoming water vapor and the resulting syngas produced by the gasifier, respectively. Furthermore, the input exergy, which is a result of the heat energy delivered to the gasifier, is denoted by the term ($\dot{Ex}_Q$) [35].

$$\dot{E}_D = \dot{Ex}_B + \dot{Ex}_{steam} + \dot{Ex}_Q - \dot{Ex}_{syngas} \quad (4)$$

Biomass possesses a negligible amount of physical exergy, leading to its frequent omission in total exergy calculations. The exergy associated with biomass, which comprises solely chemical exergy, is expressed through Eq. 5. This separation plays a crucial role by allowing for a more straightforward analysis that emphasizes the predominant aspect of biomass exergy: its chemical potential. By disregarding the minimal contribution of physical exergy, researchers can optimize their evaluations, directing their attention to the elements that have a more significant impact on energy production and the sustainability of biomass applications [36].

$$\dot{E}x_B = \dot{n}_B\left(\beta.LHV_B\right) \tag{5}$$

In connection with Eq. 5, the lower heating value for biomass fuel is employed, while the coefficient β is established as the proportion of hydrogen to carbon and oxygen to carbon. This figure is derived using Eq. 6 [37]. Grasping these variables is crucial due to their impact on the combustion efficiency and overall functioning of biomass as a sustainable energy resource.

$$\beta = \frac{1.0414 + \left(0.0177 \times \frac{H}{C}\right)}{1 - \left(0.4021 \times \frac{O}{C}\right)} - \frac{\left(\left(0.3328 \times \frac{O}{C}\right) \times \left(1 + \left(0.0537 \times \frac{H}{C}\right)\right)\right)}{1 - \left(0.4021 \times \frac{O}{C}\right)} \tag{6}$$

6. Conclusions

In this chapter, a wide range of valuable products that can be obtained from biomass is investigated. First, biofuels such as bioethanol, biodiesel, and biogas, detailing how they are produced from various biomass raw materials, including agricultural residues, forest products, and energy crops, are discussed. Following this, the topic of waste conversion into energy, along with methods of energy extraction from biomass, is examined. Research indicates that in general, thermochemical processes—particularly gasification—exhibit faster conversion rates and higher efficiency compared to other methods. One of the main advantages of biomass gasification over other thermochemical conversion technologies is its ability to operate at very high temperatures. These elevated temperatures lead to the production of syngas with improved quality and greater purity, as unwanted impurities are effectively removed. The primary product of biomass gasification is a mixture of combustible compounds known as producer gas. The composition and calorific value of this gas mixture depend on various factors, including the type of gasifying agent used.

Acknowledgements

The author utilized AI tools to refine grammar, improve clarity, and ensure accuracy of the text.

Conflict of interest

The authors declare no conflict of interest.

Biomass to Energy: Harnessing Nature's Potential
DOI: http://dx.doi.org/10.5772/intechopen.1008436

Author details

Mohammad Ali Sabbaghi
Yazd University, Yazd, Iran

*Address all correspondence to: m.sabbaghi@yazd.ac.ir

References

[1] Sabbaghi MA, Soltani M, Rosen MA. A comprehensive 6E analysis of a novel multigeneration system powered by solar-biomass energies. Energy. 2024;**297**:131209. DOI: 10.1016/j.energy.2024.131209

[2] Bridgwater T. Biomass for Energy. Journal of the Science of Food and Agriculture. 2006;**86**:1755-1768

[3] Sabbaghi MA. Enhanced exergy analysis of a solar-biomass hybrid system for micro-scale combined heat and power generation. International Journal of Exergy. 2024;**43**(4):362-374

[4] Quevedo RA, Blanca A, Escalera P, Aida V, Arias MR, Reynel HE, et al. Application of Waste Biomass for the Production of Biofuels and Catalysts: A Review. Berlin Heidelberg: Springer; 2024

[5] Sabbaghi M, Sefid M. Exergy-environmental analysis and optimization of the modified organic Rankine cycle driven by geothermal and biomass energies. International Journal of Exergy. 2023;**40**(2):144-161

[6] Lepage T, Kammoun M, Schmetz Q, Richel A. Biomass-to-hydrogen: A review of main routes production, processes evaluation and techno-economical assessment. Biomass and Bioenergy. 2021;**144**:105920

[7] Sabbaghi MA, Sefid M. Risk, sensitive analysis, and optimization of a new multi-generation system activated by solar and biomass energies. Energy Reports. 2024;**11**:1006-1018

[8] Gomez-Barea A, Leckner B. Modeling of biomass gasification in fluidized bed. Progress in Energy and Combustion Science. 2010;**36**:444-509. DOI: 10.1016/j.pecs.2009.12.002

[9] Sabbaghi M, Sefid M. Thermo-economic analysis of a combined flash-binary geothermal cycle with a transcritical CO_2 cycle to produce green hydrogen. International Journal of Exergy. 2024;**43**(2):110-122

[10] Ibrahim A, Veremieiev S, Gaskell PH. An advanced, comprehensive thermochemical equilibrium model of a downdraft biomass gasifier. Renewable Energy. 2022;**194**:912-925

[11] Xing L, Li J. Proposal of biomass/geothermal hybrid driven polygeneration plant centering cooling, heating, power, and hydrogen production with CO_2 capturing: Design and 3E evaluation. Fuel. 2022;**330**:125593

[12] Lui J, Chen WH, Tsang DCW, You S. A critical review on the principles, applications, and challenges of waste-to-hydrogen technologies. Renewable and Sustainable Energy Reviews. 2020;**134**:110365

[13] Wang Z, He T, Qin J, Wu J, Li J, Zi Z, et al. Gasification of biomass with oxygenenriched air in a pilot scale two-stage gasifier. Fuel. 2015;**150**:386-393

[14] Rahman MM, Henriksen UB, Ciolkosz D. Startup process, safety and risk assessment of biomass gasification for off-grid rural electrification. Scientific Reports. 2023;**13**:21395

[15] Susastriawan AA, Purwanto Y. Biomass gasifier-internal combustion engine system: Review of literature. International Journal of Sustainable Engineering. 2020;**14**:1090-1100

[16] Sabbaghi M, Sefid M. Evaluation of a modified organic Rankine cycle to produce power, hydrogen, and desalinated water by combining geothermal and biomass energies from 4E viewpoints. Journal of Mechanical Engineering University of Tabriz. 2022;**52**(4):301-310

[17] Gírio F, Catarino J. Sustainable value methodology to compare the performance of conversion technologies for the production of electricity and heat, energy vectors and biofuels from waste biomass. Cleaner Waste Systems. 2022;**3**:100029

[18] Couto N, Silva V, Monteiro E, Teixeira S, Chacartegui R, Bouziane K, et al. Numerical and experimental analysis of municipal solid wastes gasification process. Applied Thermal Engineering. 2015;**78**:185-195

[19] Kumar A, Parihar S, Hammer T, Sridhar G. Development and testing of tube type wet ESP for the removal of particulate matter and tar from producer gas. Renewable Energy. 2015;**74**:875-883

[20] Gao N, Salisu J, Quan C, Williams P. Modified nickel-based catalysts for improved steam reforming of biomass tar: A critical review. Renewable and Sustainable Energy Reviews. 2021;**145**:111023

[21] Sabbaghi MA, Sefid M. Energy, exergy, economic and environmental analysis (4E) of an organic Rankine cycle to produce power, hydrogen and desalinated water by combining geothermal energy and heat recovery. Journal of Solid and Fluid Mechanics. 2022;**12**(5):201-214. DOI: 10.22044/jsfm.2023.11259.3477

[22] Valderrama Rios ML, González AM, Lora EES, Almazán del Olmo OA. Reduction of tar generated during biomass gasification: A review. Biomass and Bioenergy. 2018;**108**:345-370

[23] Sabbaghi MA, Baniasadi E, Genceli H. Thermodynamic assessment of an innovative biomass-driven system for generating power, heat, and hydrogen. International Journal of Hydrogen Energy. 2024;**75**:529-539. DOI: 10.1016/j.ijhydene.2024.03.196

[24] Pio DT, Tarelho LACC, Pinto RG, Matos MAAA, Frade JR, Yaremchenko A, et al. Low-cost catalysts for in-situ improvement of producer gas quality during direct gasification of biomass. Energy. 2018;**165**:442-454

[25] Deepanraj SR. Recent advances in direct catalytic thermochemical gasification of biomass to biofuels. In: Verma P, editor. Thermochemical and Catalytic Conversion Technologies for Future Biorefineries. Vol. 1. Springer Nature Singapore: Singapore; 2022. pp. 241-295

[26] Suparmin P, Nurhasanah R, Hendri H. Biomass for dual-fuel syngas diesel power plants. Part I: The effect of preheating on characteristics of the syngas gasification of municipal solid waste and wood pellets. Arab Journal of Basic and Applied Sciences. 2023;**30**:378-392

[27] Richardson Y, Drobek M, Julbe A, Blin J, Pinta F. Biomass Gasification to Produce Syngas. Recent Advances in Thermo-Chemical Conversion of Biomass. Elsevier; 2015

[28] Abdoulmoumine N, Adhikari S, Kulkarni A, Chattanathan S. A review on biomass gasification syngas cleanup. Applied Energy. 2015;**155**:294-307

[29] Yeo TY, Ashok J, Kawi S. Recent developments in sulphur-resilient catalytic systems for syngas production.

Renewable and Sustainable Energy Reviews. 2019;**100**:52-70

[30] Huang MT, Zhai PM. Achieving Paris agreement temperature goals requires carbon neutrality by middle century with far-reaching transitions in the whole society. Advances in Climate Change Research. 2021;**12**:281-286

[31] Sarić M, Dijkstra JW, Haije WG. Economic perspectives of power-to-gas technologies in bio-methane production. Journal of CO_2 Utilization. 2017;**20**:81-90

[32] Prabakaran G, Vigneswaran VS, Gowd SC, Mishra R, Singh E, Kumar A, et al. Bioenergy with carbon capture, storage and utilization: Potential technologies to mitigate climate change. Biomass Bioenergy. 2023;**177**:106941

[33] Zhu J, Su Y, Chai J, Muravev V, Kosinov N, Hensen EJM. Mechanism and nature of active sites for methanol synthesis from CO/CO_2 on Cu/CeO_2. ACS Catalysis. 2020;**10**:11532-11544

[34] Loha C, Chatterjee PK, Chattopadhyay H. Performance of fluidized bed steam gasification of biomass – Modeling and experiment. Energy Conversion and Management. 2011;**52**:1583-1588

[35] Ptasinski KJ, Prins MJ, Pierik A. Exergetic evaluation of biomass gasification. Energy. 2007;**32**:568-574

[36] Pellegrini LF, de Oliveira JS. Exergy analysis of sugarcane bagasse gasification. Energy. 2007;**32**:314-327

[37] Sabbaghi MA, Soltani M, Fraser R, Dusseault MB. Emergy-based exergoeconomic and exergoenvironmental assessment of a novel CCHP system integrated with PEME and PEMFC for a residential building. Energy. 2024;**305**:132301

Chapter 3

Perspective Chapter: The Growing Importance of Bio-Attributed Products in Reducing Greenhouse Gas Emissions

Ivana Krkljuš, Peter Saling and Christian Krüger

Abstract

Renewable raw materials, as regenerative biobased feedstock for chemicals, have an enormous potential for reducing feedstock-related greenhouse gas emissions. Dedicated to driving the growth of bioeconomy, we continuously explore and develop building blocks for turning the bioeconomy ambition into a reality. Bio-based products are already available in the market but are limited in growth due to lack of technologies and product development costs. Alternatively, scale-up and utilization of biobased feedstocks, leveraging existing infrastructure, requires utilization of different chain-of-custody models. That is how a growing need for bio-attributed products, which provide the same quality and performance of their conventional equivalent and a proven benefit determined by the reduction of the Product Carbon Footprint in cradle-to-gate assessments, can be fulfilled. Ranging from a discussion around collaboration to drive sustainability in the chemical industry, towards current and future policy asks, we discuss about the opportunities of using different types of feedstocks and how we can accelerate industry transformation strategies by using attribution models, model biogenic CO_2-uptake in a meaningful, fair, and transparent way, based on accepted and harmonized standards, such as global ISO standards. We provide reference points and statistics and address the latest and the most important bioeconomy-related aspects of bio-attribution as science is evolving.

Keywords: bioeconomy, bio-attributed products, biogenic carbon, mass balance chain of custody, GHG emissions reduction

1. Introduction

Reducing carbon emissions is crucial for limiting global warming. Fossil fuels are a major source of greenhouse gases, which contribute to climate change. By moving away from fossil fuels, we can reduce carbon footprint and mitigate environmental damage.

There are two major concepts essential for achieving climate neutrality: de-fossilization and decarbonization, focused on different aspects of energy and industrial systems. *De-fossilization* refers to the process of reducing or eliminating the use of fossil resources, such as coal, oil, and natural gas, for energy and materials production. *Decarbonization* focuses on reducing the amount of carbon dioxide (CO_2) and other carbon compounds released into the atmosphere from human activities. It is a broader concept that includes de-fossilization, but also encompasses other strategies to lower carbon emissions.

For chemical industry, it is essential to use carbon because it is a fundamental building block for many chemical compounds necessary for our daily life. Carbon is incredibly versatile and can form strong bonds with each other and with atoms, allowing it to create complex structures. These structures determine the properties and functions of chemicals. Without carbon, the chemical industry would not be able to produce the materials and compounds that are crucial for everyday products and technologies.

The chemical industry is a cornerstone of a modern economy that supports the value chains of various sectors and plays a pivotal role in driving the sustainability of all downstream industries. Not only does it provide employment to 1.2 million people in Europe [1], but it also serves as a catalyst for innovation. The significance of a strong and resilient European chemical industry cannot be emphasized enough. The industry is currently undergoing a significant transformation to align with the ambitious objectives of the European Green Deal [2], which aims to achieve climate neutrality, circularity, and digitalization. This transformation requires substantial investments amounting to many billions of euros over the coming decades, leading up to the 2050 deadline. A robust European chemical industry is vital for the successful implementation of the European Green Deal, as chemicals are present in more than 90% of manufactured goods, and almost everything from solar panels to pharmaceuticals is made with the help of chemicals [3], serving as a foundation for modern lifestyles and the advancement of clean technologies.

Today, the chemical industry mainly relies on fossil resources such as oil, coal, and natural gas, with petrochemicals making up a significant portion of its raw materials. In 2019, fossil-based feedstocks accounted for around 80% of the total feedstock used [4]. The phaseout of fossil resources is an urgent need [5], and the chemical industry has a critical role to play in this transition [6]. De-fossilization recognizes the need to transition away from these finite resources toward renewable and recycled feedstocks, such as bio-based or recycled materials, and CO_2, **Figure 1**, as well as adopting solar, and wind as energy sources, or hydrogen to power manufacturing processes.

The importance of utilizing renewable and recycled feedstock becomes even more apparent when consulting the following statistics.

Share of Fossil Carbon: In 2020, approximately 99% of the raw materials used for plastics production in the EU were derived from fossil fuels, including oil and natural gas [7]. Globally, over 95% of the raw materials used in plastic production come from fossil sources [8]. Fossil fuels accounted for 84.3% of the global primary energy consumption in 2019 [9].

Share of Recycled and Renewable (Circular) Carbon: The recycling rate for plastic packaging was raised up to 39.7% in 2021 [10]. Globally, only about 9% of plastic waste is recycled. The rest ends up in landfills, is being incinerated, or leaks into the environment [11]. The share of circular carbon embedded in products is still low, compared to the amount of fossil carbon. Kähler et al. have published a share of 9% of embedded circular carbon in chemicals and derived materials globally [12]. The global steel industry recycles 33% of the total steel production annually [13].

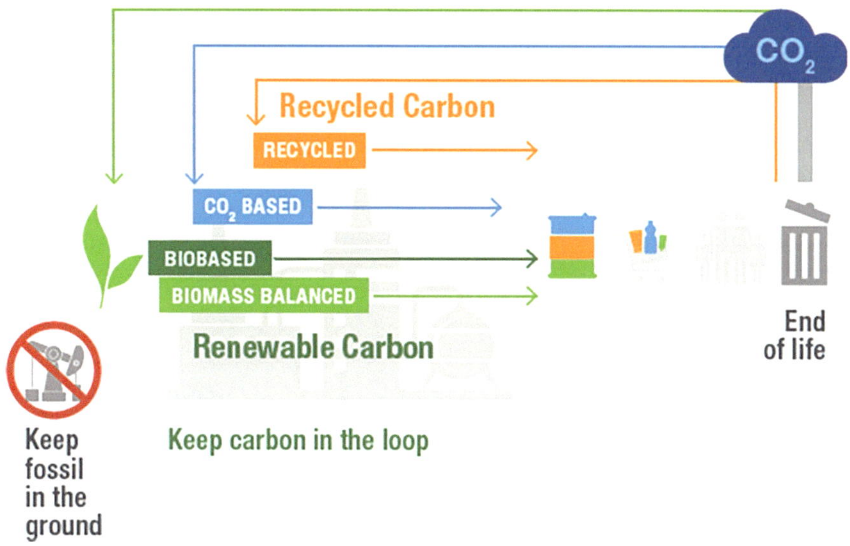

Figure 1.
Renewable carbon sources replacing fossil resources.

Share of Biobased Products: The share of bio-based products in the EU's chemical industry was around 8% in 2017 [14]. The global production capacity of bioplastics is accounting for approximately 1% of the total plastics market [15].

Obviously, we need other concepts beyond bio-based products to enable transformation [16]. Bio-attributed products offer a new paradigm for sustainable solutions by combining renewable and traditional resources, while leveraging the existing infrastructure to deliver affordable products without compromising performance, i.e. through adoption of mass-balanced products [17]. Mass balance allows for a gradual integration of renewable materials into existing production systems, without the need for separate production lines. The emphasis is on substituting the overall fossil carbon demand for a customer product, rather than tracing and substituting carbon from each individual input in a single product value chain. This provides greater flexibility and allows for the substitution of more fossil carbon [18].

Bio-attributed products benefit from greenhouse gas emissions reduction via biogenic carbon[1], i.e. assimilation of CO_2 from the atmosphere at the farming step. The carbon that is released into the atmosphere when biogenic materials are burned or decayed is offset by the carbon that is absorbed by the plants or trees. In contrast, the consumption of fossil resources releases carbon that has been sequestered underground for millions of years, contributing to the increase of carbon dioxide in the atmosphere and exacerbating climate change.

Yet, carbon accounting methodologies for bio-derived products (aggregated term for bio-based and bio-attributed products) require an advancement in carbon

[1] Biogenic carbon refers to carbon that is sequestered from the atmosphere during the growth of biomass. This carbon is found in a variety of natural materials, such as trees, plants and other forms of biomass, and accumulates in reservoirs as organic carbon in the soil.
Biogenic CO2 emissions are defined as CO2 emissions related to the natural carbon cycle, as well as those resulting from the combustion, harvest, digestion, fermentation, decomposition, or processing of biologically based materials.

accounting standards, to ensure a comprehensive and accurate assessment of the climate impact of bio-derived products throughout their life cycle.

2. Prerequisites for the successful transformation towards a sustainable future

To successfully establish a transformation pathway, there are several important prerequisites that need to be addressed.

1. *Exploring the bioeconomy potential*

   Sections 3–4 delve into the concept of bioeconomy, highlighting the importance of utilizing renewable resources.

2. *Demonstrating significant progress in reducing GHG emissions*

   Sections 5–8 highlight the importance of rewarding concepts and Chain-of-custody (CoC) models that enable significant emission reduction measures early on, as well as supporting progress in sustainable management of complex value chains.

3. *Integrated assessment methodologies and frameworks*

   In Section 9–10 we present the integration of a Mass Balance Chain-of-custody model into standardized Life Cycle Assessment (LCA) to achieve accurate and comprehensive sustainability assessments. We will also delve into the application of a specific LCA indicator, the Product Carbon Footprint (PCF), to further highlight the benefits of this integrated approach.
   By incorporating the holistic and consistent approach to −1/+1 biogenic carbon accounting,[2] businesses can comprehensively assess and communicate the environmental impact of products throughout their life cycle, enabling informed decision-making and effective product placement strategies, Section 11.

4. *Collaboration with stakeholders and industry oriented policy*

   Section 12 highlights the role of policy and advocacy in providing recognition and incentives for sustainable practices.

3. From waste to wealth: The circular bioeconomy driving Europe's climate-neutral future

Bioeconomy [19] embraces the use of renewable resources, the valorization of waste, and the cascading use of resources [20–22]. These resources are harnessed to

[2] The "-1/+1 approach" considers all the biogenic carbon flows of a product throughout its entire life cycle. When a bio-based material is used, the biogenic carbon stored in the biomass enters the product's system and is reported as a negative emission. A positive emission is accounted for at the end of the product's useful life when the biogenic carbon is released back into the atmosphere or subsequently transferred to another product system in short time periods.

produce a wide range of essential products including food, energy, and materials like plastics, textiles, and chemicals. Additionally, bioeconomy plays a vital role in waste management [23–25].

4. Unleashing the potential: Understanding the concept of biogenic carbon

Biogenic carbon [26] refers to carbon that comes from natural living sources like plants, trees, animals, microorganisms, and another biomass. Living organisms play a crucial role in absorbing carbon dioxide from the atmosphere through photosynthesis. Unlike fossil carbon, which is stored underground for millions of years in the form of coal, oil, and natural gas, biogenic carbon is part of the Earth's current carbon cycle.

Plants absorb carbon dioxide (CO_2) from the air during photosynthesis. They use carbon to grow and produce biomass, like leaves, stems, and roots. When animals eat plants, they consume the carbon stored in the plant biomass. This carbon becomes part of their bodies. When plants and animals die, microorganisms break down their biomass, releasing carbon back into the air as CO_2 or into the soil. This process repeats continuously, making biogenic carbon part of a renewable cycle that's constantly being reused in nature.

5. The bio-attributed products: Driving sustainability and the transition to a low-carbon economy

For consumers seeking a measurable biogenic content, bio-based products [27] exist for multiple applications. The products are derived from dedicated production routes, using materials primarily from vegetable oils like palm oil, palm kernel oil, and their respective derivatives, as well as fats, grains, sugar, and wood. The bio-based content can be tracked to the final product by analytical measurements, and determined by various radiocarbon methods, e.g., ASTM D6866-12:2012 [28], CEN/TS 16640:2014 [29], DIN-51637:2014 [30].

For consumers who support reduction of fossil feedstock utilization, the fast-emerging bio-attributed products and the biomass balance concept come into play. In this approach, renewable raw materials are co-fed into the production chain alongside fossil raw materials and attributed to end products. Bio-attributed come with the same quality, performance, and properties as their fossil twins.

To support and accelerate the transition of the chemical industry, it is essential to utilize both concepts and offer both bio-based products and bio-attributed products to customers and consumers. This distinction extends to sustainability product claims that are increasingly guided by regulatory guardrails for consumer empowerment [31] and green claims [32]. Product claims refer to the marketing statements that companies make about their products, to inform consumers about the sustainability aspects of a product. Manufacturers must not claim bio-based C14 carbon content [28–30] in their bio-attributed products. Product claims of bio-attributed products may highlight the positive contribution of feedstock replacement in the value chain that goes without compromising on the product performance, compared to the fossil equivalent, reduced product carbon footprint, due to biogenic carbon utilization, as well as substitution of fossil carbon by renewable carbon.

5.1 Examples of claims

"X% attributed renewable raw material": This claim indicates that the raw material needed to manufacture the product has been substituted by renewable feedstock and this feedstock is attributed to the product via the mass balance method.

"Certified sustainable": This means that the product has been audited by a third-party organization to meet the criteria of a certification scheme, e.g. on traceability and sustainability.

6. Tracing sustainability: Understanding the different chain-of-custody models for a transparent and traceable supply chain

There are different *chain-of-custody (CoC) models* [33] in place, which offer different levels of traceability and transparency of raw materials flow throughout the supply chain. ISO 22095:2020 [33] is a global and horizontal standard that outlines the requirements for implementing CoC models. The CoC models are differentiated by the physical presence of characteristics in the final product, **Figure 2**. Each of these CoC models is suitable for certain supply-chain and production settings.

The identity preserved (IP) CoC model is used for systems where the identity and integrity of a specific product or material are maintained throughout the supply chain in industries such as food, agriculture, and specialty products. It is sometimes applied to food value chains, e.g. the material flow is followed from the specific farm or forest, where the raw materials were grown or harvested, through the processing and manufacturing stages, to the final product. This material flow is physically kept completely separated from other material flows originating from other farms or forests.

The concept of a *Segregated CoC model* generally refers to a system where materials or products from different sources or certified origins are kept separate

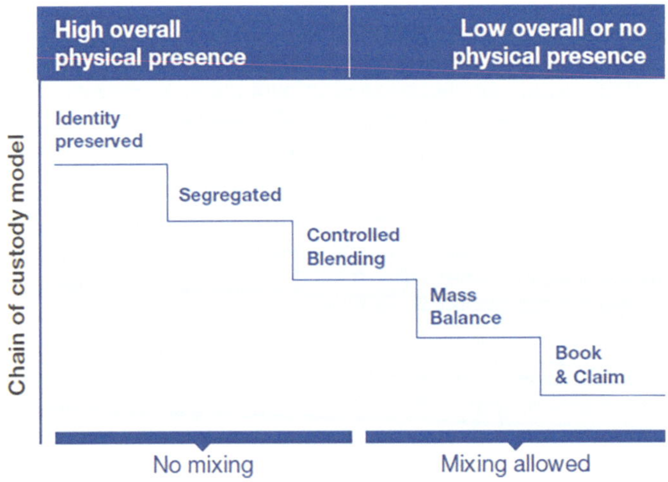

Figure 2.
Different chain-of-custody (CoC) methods, differentiated by physical presence, based on the ISO 22095:2020 [33].

throughout the supply chain, to maintain their distinct identity and ensure traceability. For renewable and fossil raw materials the segregated CoC model requires complete separation throughout the linear value chain. E.g. the 100% renewable content claim can be used if renewable content is not mixed with fossil content at any point.

Controlled blending CoC model [33] refers to a system or process that ensures the traceability and integrity of materials being blended in a controlled manner at any time. This concept is commonly used in industries such as forestry, agriculture, and manufacturing, where multiple materials or components are combined to create a final product. Typical product examples of bio-based products are products where the biogenic part is "chemically mixed" with a non-biogenic part. The share of biogenic part is known at any point of time throughout the supply chain and can be analytically measured [28–30].

For materials, in a complex production environment (e.g. multi-input, multi-output production set-up), today the most relevant attributional CoC model is the *Mass Balance (MB) model*, **Figure 3**. MB allows to simultaneously use a mix of renewable and conventional feedstocks, using the existing interlinked/integrated system of production plants, energy flows, and infrastructure. The use of renewable carbon can be scaled up effectively and rapidly without costly adaptation of production processes and value chains. A third party certification ensures the credibility of the system by avoiding double counting.

The Book and Claim (B&C) CoC model [33] may be an option for sharing sustainability attributes between companies without physical transport of energy, or in case of materials without physical transport of products. B&C systems are the most applied method for tracking energy attributes because Energy Attribute Certificates

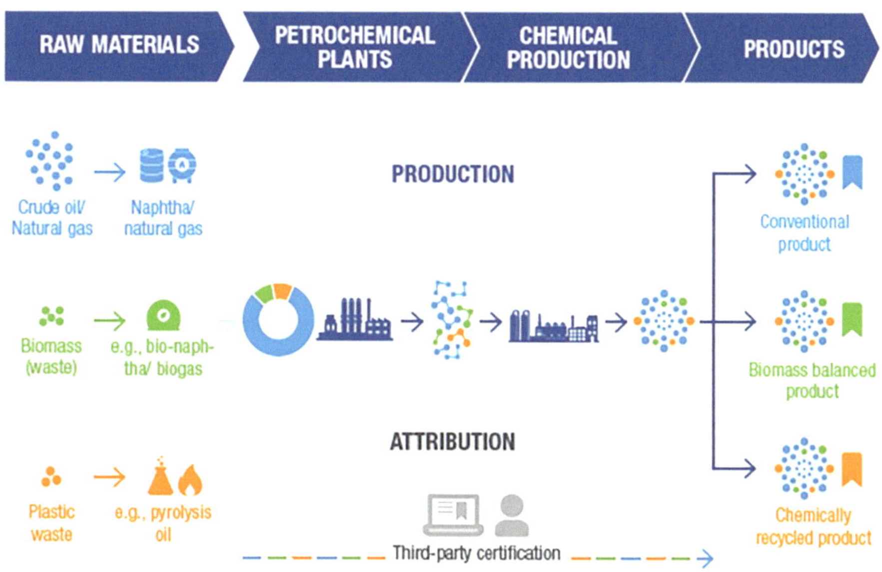

Figure 3.
The mass balance approach in chemical production. One or more characteristics (e.g. biogenic carbon) are attributed to certain physical outputs, i.e. products.

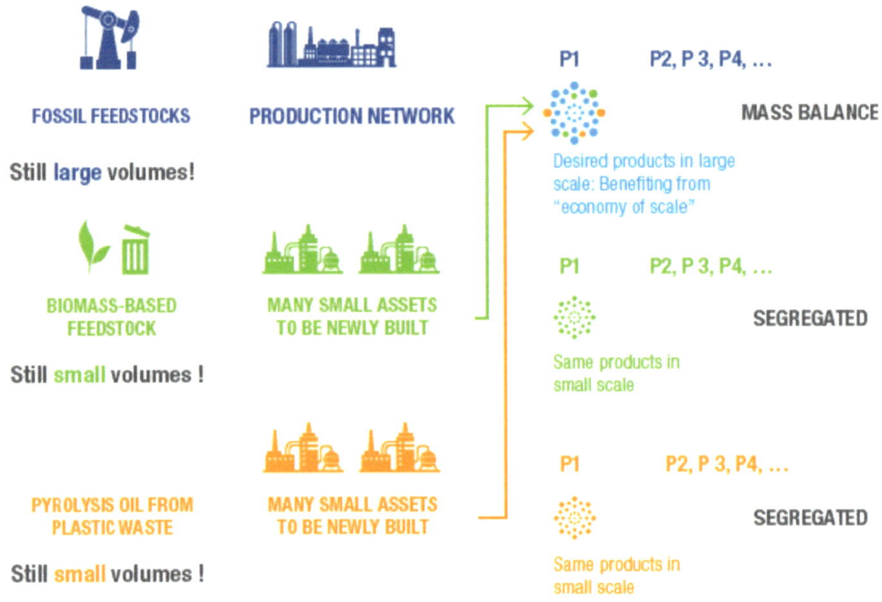

Figure 4.
CoC models and their application to generate different types of products.

(EAC)[3] [34], such as GOs,[4] can be transferred and traded separately from the physical flow of energy. Every unit of energy generated has certain characteristics, such as the source of the energy (wind, solar, coal), where the energy was generated, and when the production device first came online. These characteristics can be documented into an EAC which is issued to the producer and can be traded or transferred to a consumer. Because there is informational and economic value to EACs, they must be accurate and securely managed. Energy attribute tracking systems based on the B&C methodology exist across the world.

Figure 4 gives a general overview of the segregated and mass balance CoC models and their direct impact on final products.

The IP, Segregated and Controlled blending CoC imply a measurable physical content, while attributional models like MB and B&C imply an attributed content, which is not measurable [34].

---

[3] Energy Attribute Certificates (EACs) are the vehicle used to carry energy attributes certified via an attribute tracking system. In Europe, the primary certificate used by attribute tracking systems is the Guarantee of Origin or "GO". In other locations, like the United States, the certificate used is the REC (Renewable Energy Certificate). An EAC is often bought, sold and canceled with prices determined by a supply and demand market.

[4] The Guarantee of Origin (GO or GoO) is the tracking certificate regulated by the EU Renewable Energy Directive 2023/2413 (RED-3 [34]). The GO is the carrier of energy attributes. The trade, cancelation and use of GOs are further governed by the European Standard CEN EN 16325 [35], as referenced in the RED-3. In addition, the European Energy Certificate System (EECS) rules, maintained and enforced by the Association of Issuing Bodies (AIB) provides further rules that serve to harmonize and standardize GO markets. GO prices are determined by a voluntary the supply and demand market.

7. Balancing sustainability: Unveiling the power of the mass balance approach in a circular economy

The mass balance approach is already well established in biofuel markets. Both green methane (from renewable sources) - biomethane, and gray methane (from conventional sources) - natural gas, can coexist within the same network or pipeline grid. This approach allows for mixing and blending of different sources of methane in a common infrastructure. The consumer receives gas from the shared network, without being able to differentiate between the different sources. The mass balance model ensures that the total amount of renewable or green gas, injected into the network, matches the amount consumed by users, even if the specific molecules from renewable sources cannot be physically traced to the end consumer.

The mass balance model is also applied to track the flow of materials through complex production systems in the chemical industry, which is designed to create a wide range of products, including polymers, coatings, crop-protection agents, vitamins, and more [36], Video 1 (https://www.youtube.com/watch?v=7vZa36ikBi8).

The mass balance model enables the substitution of hydrocarbons from fossil sources with hydrocarbons from non-fossil sources at the beginning of petrochemical value chain (input) and is attributed to the product (output) in such a manner that the input and the output match, i.e., the overall balance of raw material attributes is maintained. The recycled and renewable feedstock demand is calculated back to the primary carbon sources, like naphtha and natural gas, which are replaced with an adequate volume of biogas, bio-naphtha, or pyrolysis oil (recycled from post-consumer or pre-consumer waste). The following prerequisites must be fulfilled:

1. accurate consideration of all raw materials inputs,

2. tracing raw materials and characteristics flows,

3. maintaining the overall balance of raw materials and products, ensuring the sum of input equals the sum of output, via predefined rules or calculations,

4. accounting for process losses/gains (e.g., the fossil fuels, utilized for steam and electricity generation or fuel production),[5]

5. the mathematical attribution of certified content of raw materials (input) to specific products (output).

The traceability[6] and the correct balance between inputs and outputs [33] can be ensured by audits, according to a set of transparent rules and a third-party certification approach. The most common certification schemes in the chemical sector are the REDcert[2] scheme [37] and the ISCC PLUS [38]. Certification schemes involve the evaluation of specific criteria established by horizontal or sector-specific standards, by independent auditors. The availability of transparent and accurate product claims, based on certificates, empowers downstream customers and consumers to make choices that align with their sustainability preferences.

[5] The same applies for inorganic intermediates like sodium hydroxide or chloric acid.

[6] Traceability is defined according to ISO 22095: ability to trace the history, application, location or source(s) of a material or product throughout the supply chain [33].

The following guidelines are outlined in the ISO 22095:2020 [33]:

1. Documentation and record-keeping: The standard emphasizes the importance of maintaining accurate and comprehensive documentation throughout considered system boundaries of different CoC models, including the production networks. This includes recording the origin, quantity, and characteristics of the input materials, with characteristics mixed with conventional inputs (blended process), the processes to produce an output, and the flow of characteristics, as well as any relevant information about sustainability or quality standards.

2. Material identification: ISO 22095:2020 [33] highlights the need for clear identification of different raw materials, to maintain their traceability.

3. Verification and certification: The standard encourages companies to obtain third-party certifications or execute audits, to provide independent verification of their CoC systems.

4. Training and competence: The personnel involved in the application of the different CoC models must have a solid understanding the CoC requirements, proper handling and storage of materials, and adhere to relevant procedures and documentation.

5. Internal audits and management review: The standard recommends conducting regular internal audits to identify any non-conformities, implement corrective actions, and continuously improve the certification process.

8. The two implementing methods of mass balance chain-of-custody model

The mass balance method considers the flow of characteristics within the defined boundaries and ensures that the total quantity or value of the characteristics entering the system matches the quantity or value of those leaving the system. Within the system boundaries, the sum of the physical input and the output mass flows must be identical.

In principle, there are two different implementation methods according to ISO 22095 [33]. *The rolling average mass balance method*, **Figure 5**, is a specific approach used in the mass balance system to account for the mixture of inputs with and without specific characteristics. This method involves considering the average characteristics of inputs over a defined period, typically 1 year. In the context of bio-based renewable raw materials, the rolling average method allows for the combination of sustainably certified and non-certified bio-based renewable raw materials. The average of inputs with certified bio-based characteristics is determined over the defined booking period, and this average is then attributed to all outputs in the same output ratio.

The credit method mass balance, **Figure 6**, is a specific approach within the mass balance system that involves generating credits based on specific input characteristics, such as the renewable attribute of a renewable feedstock. These credits are then stored in a credit account and can be used to balance the use of certain input characteristics for the attribution on final products. By utilizing the credit method organizations using mass balance can demonstrate the use of renewable or other desired attributes in their products.

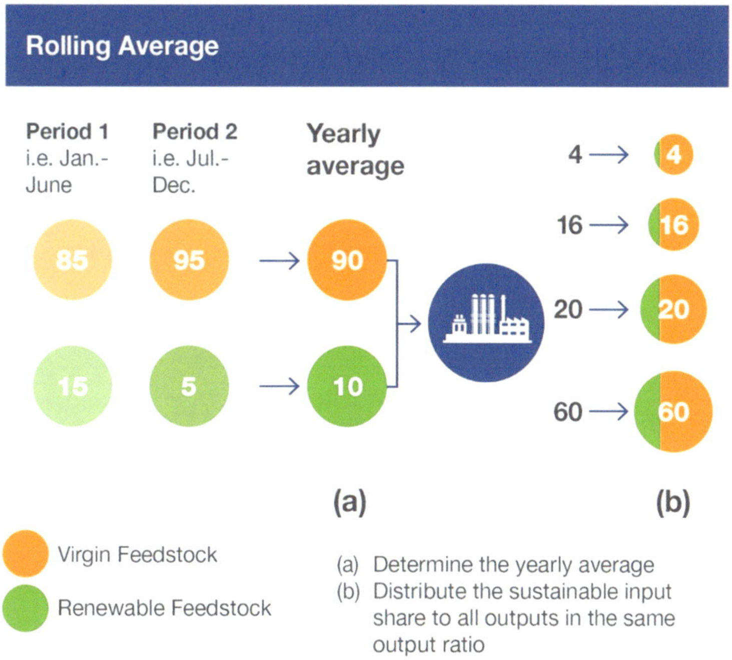

Figure 5.
Illustration of the rolling average implementation method of MB CoC model.

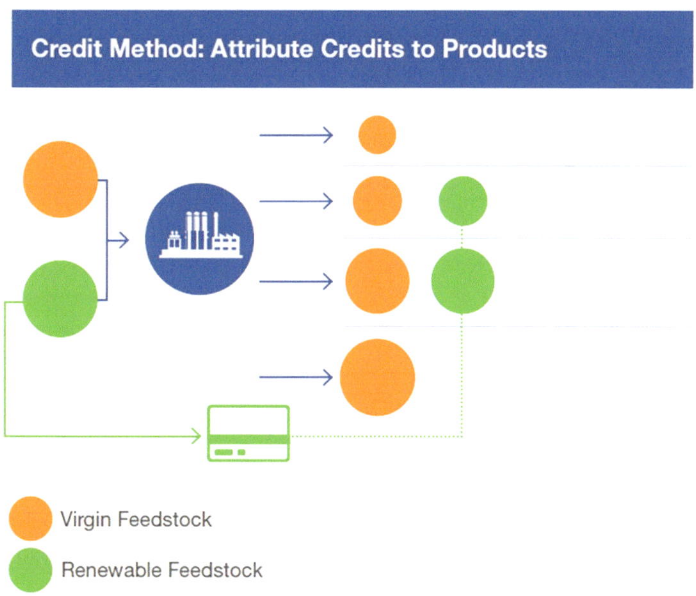

Figure 6.
Illustration of the credit method of MB CoC model.

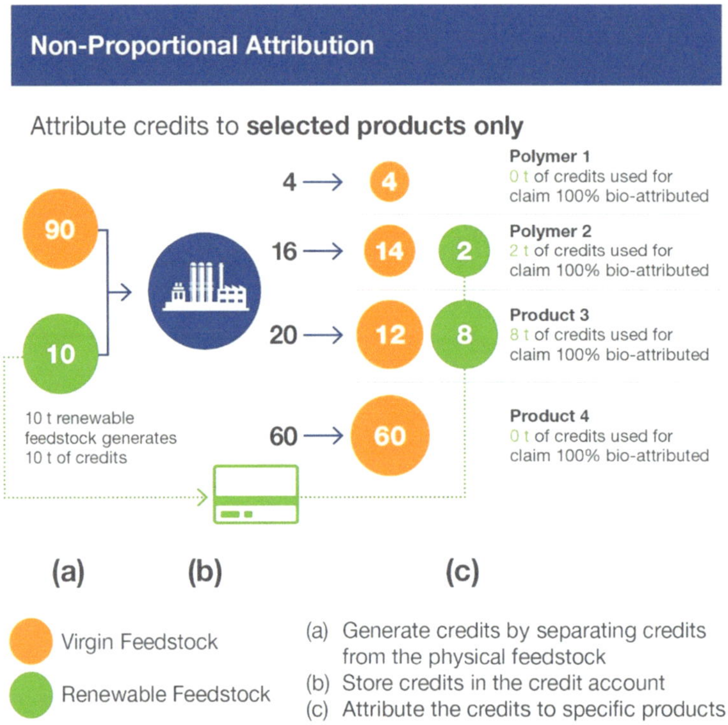

Figure 7.
Illustration of the non-proportional attribution of the renewable characteristics of credit MB CoC for multi-output processes within a system boundary.

ISO/CD 13662 [39][7] provides guidance on the application of credit mass balance to multi-output processes. In the case of *non-proportional attribution*, **Figure 7**, the credits generated from a specific input characteristic, such as renewable content, are attributed to the different outputs of a multi-output process, based on their relative contribution to the overall process. This approach allows for a targeted attribution of renewable characteristics to specific products. An example is given in **Figure 7**, where renewable characteristics are directly attributed to Polymer 2 and Product 3.

On the other hand, *the proportional attribution*, **Figure 8**, distributes the renewable characteristics based on the output yield of each product. The credits generated from a specific input characteristic are attributed to each output in direct proportion to their physical quantity or mass, i.e. in proportion to their contribution to the overall production process. In **Figure 8**, the renewable characteristics are proportionally distributed among all outputs, based on their respective yields.

The non-proportional attribution model is currently more widely used in voluntary markets. Given the existing market dynamics and the cost differentials between renewable feedstocks and their fossil counterparts, the utilization of the non-proportional attribution model helps mitigate the price difference and facilitates the sale of bio-attributed products at a competitive price point. On the other hand, certain regulated markets [40] require proportional attribution.

[7] The ISO/CD 13662 [39] is under development. A draft is being reviewed by the technical committee ISO/TC 308/WG2. The published version of the ISO/CD 13662 may contain further updates.

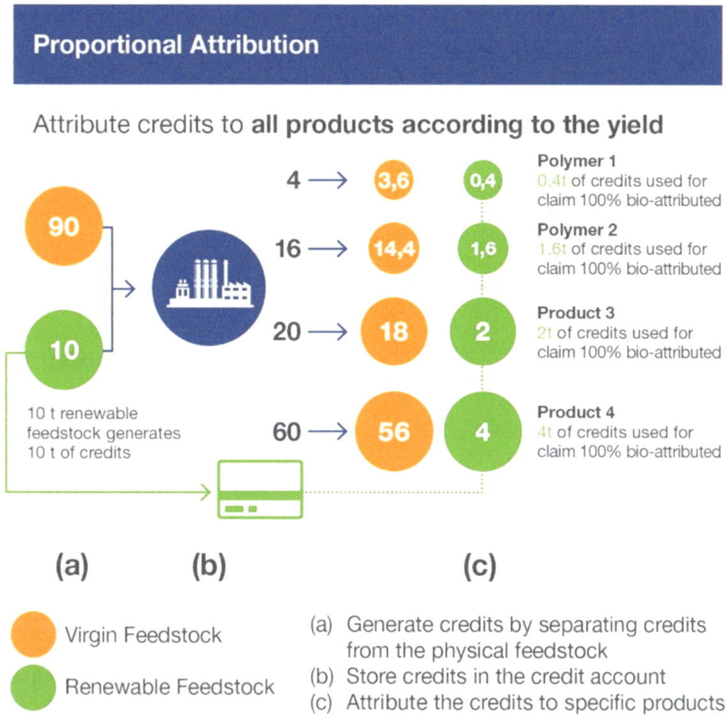

Figure 8.
Illustration of the proportional attribution of the renewable characteristics of credit MB CoC for multi-output processes within a system boundary.

9. Enhancing sustainability assessment: The synergy of mass balance chain of custody model and life cycle assessment

Life cycle assessment (LCA), as defined by ISO 14040 [41], involves the compilation and evaluation of inputs, outputs, and potential environmental impacts of a product system throughout its life cycle. LCA is a valuable tool for assessing the environmental performance of various products, e.g., bio-derived products. By considering factors such as raw material extraction, manufacturing processes, distribution, use, and disposal or recycling, LCA offers a holistic view of a product's environmental performance. Accurately accounting for the biogenic carbon uptake is crucial in LCA studies of bio-derived products. The choice of system boundaries (the limits of the LCA assessment) is essential in determining the extent to which biogenic carbon is considered in the assessment [41, 42].

To improve the accuracy, reliability, and robustness of environmental assessments, the MB CoC model can be combined with the LCA. The integration of the MB approach into LCA for biomass applications in the chemical sector follows a well-defined methodology [43]. This methodology has been critically reviewed by a panel, primarily comprising academic experts, confirming its adherence to the requirements of ISO 14040 [41] and ISO 14044 [42]. To apply this methodology several use cases involving bio-based feedstocks, specifically biomass balance, have been examined [44, 45]. In 2019 a methodology was published [43], based on a material flow analysis and the substitution of feedstock datasets. The aim was to determine the number of sustainable feedstocks required to replace fossil inputs through material flow analysis.

The life cycle inventory of outputs, which includes products with attributed sustainable content obtained through the mass balance approach, is determined based on the relative conversion rates and chemical values of the different feedstocks. Overall, the integration of concepts and methodologies of both approaches enhances the accuracy and reliability in the following manner:

1. It involves using both approaches to *track the flow of materials, energy, or products within a supply chain*, and *ensure transparency and traceability*. The MB CoC model helps to collect the required data accurately (tracks the flow of materials, energy, or products within a supply chain), which is essential for a reliable LCA.

2. The MB data help *understand where the product's inputs and outputs come from*. These accurate and reliable data flows, guaranteed by certification schemes, can then be incorporated into the LCA framework.

3. The model can also be used to *attribute specific environmental impacts*, based on the measured or estimated flow of materials or energy, to processes or products within a specific supply chain. This helps us accurately reflect the environmental burdens associated with each stage of the life cycle.

4. By incorporating the MB data into the LCA framework, *discrepancies or inconsistencies* can be identified and *the credibility of the LCA assessment enhanced*.

10. Unleashing the potential of product carbon footprint (PCF) in the context of quantifying the environmental impact of products

Within the realm of LCA, the Product Carbon Footprint (PCF) [41, 42, 46] has emerged as a widely recognized indicator of environmental impact. The PCF focuses on quantifying greenhouse gas emissions associated with a product. In 2024, Together for Sustainability (TfS) [47] published an industry-specific PCF guideline, to ensure that the assessment accurately captures the carbon footprint of chemical products [48, 49]. **Figure 9** demonstrates examples of the integration and PCF calculation of bio-attributed products, as outlined by TfS [48]: the "conventional reference" approach and the "inventory" approach. Both approaches are accepted methods for accurately calculating PCFs, with varying input shares of sustainable feedstocks indicated by numbers 1 and 2 in the figure.

To communicate the results and facilitate the comparison of environmental impacts, the ISO 14020 [50] series should be consulted. This allows for the creation of self-declared claims, ecolabels [51], Environmental Product Declarations (EPDs) [52], or other forms of footprint communication.

11. Advancing sustainability: Carbon accounting rules for valorization of renewable raw materials in the value chain

It is a common practice to consider biogenic carbon dioxide emissions from biomass-derived products as climate neutral. This is because the carbon released during the combustion or decomposition of biomass-derived products is equivalent to the carbon that was previously absorbed from the atmosphere during the growth

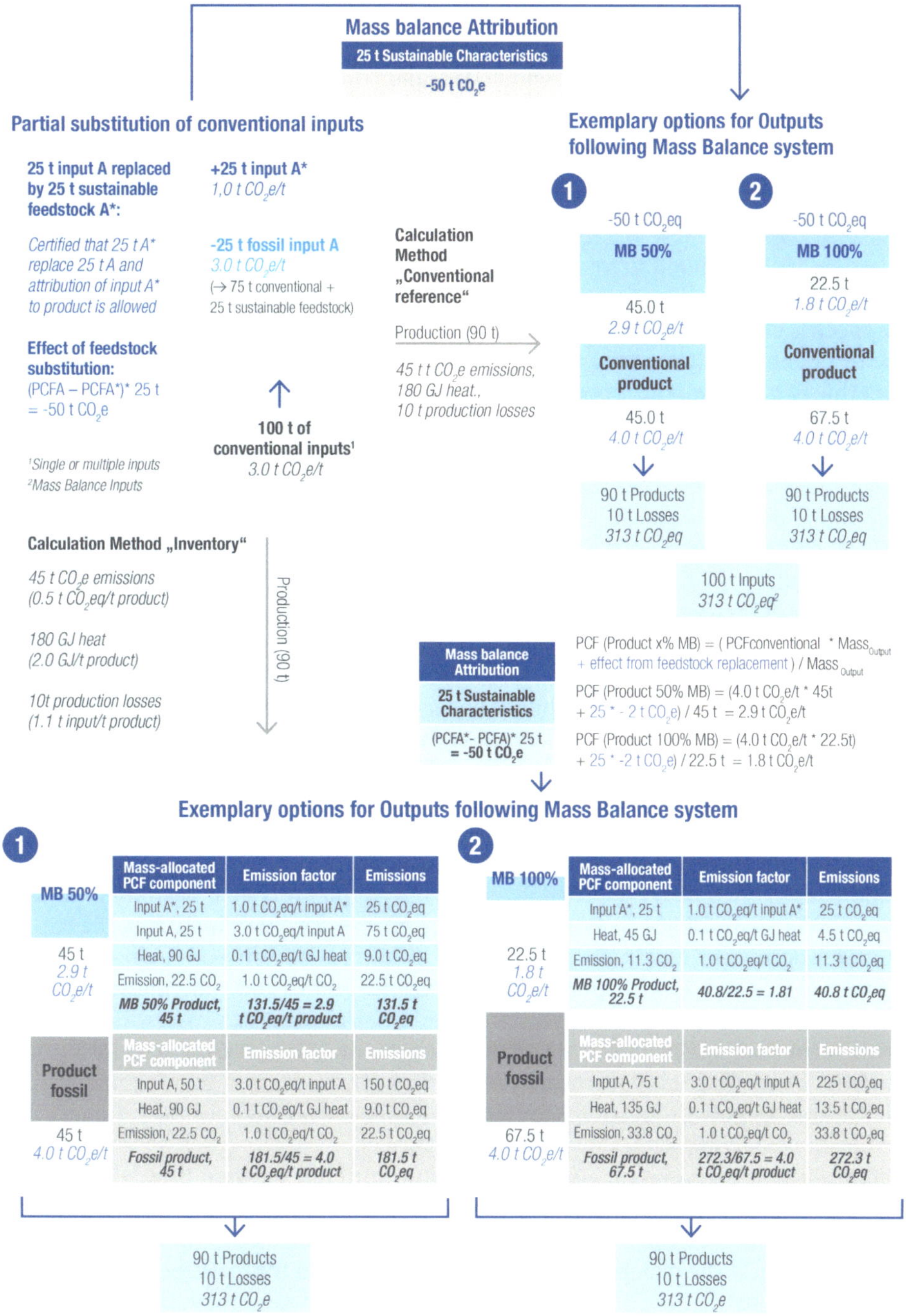

Mass-allocated PCF component	Emission factor	Emissions
Input A*, 25 t	1.0 t CO_2eq/t input A*	25 t CO_2eq
Input A, 25 t	3.0 t CO_2eq/t input A	75 t CO_2eq
Heat, 90 GJ	0.1 t CO_2eq/t GJ heat	9.0 t CO_2eq
Emission, 22.5 CO_2	1.0 t CO_2eq/t CO_2	22.5 t CO_2eq
MB 50% Product, 45 t	***131.5/45 = 2.9 t CO_2eq/t product***	***131.5 t CO_2eq***

Mass-allocated PCF component	Emission factor	Emissions
Input A, 50 t	3.0 t CO_2eq/t input A	150 t CO_2eq
Heat, 90 GJ	0.1 t CO_2eq/t GJ heat	9.0 t CO_2eq
Emission, 22.5 CO_2	1.0 t CO_2eq/t CO_2	22.5 t CO_2eq
Fossil product, 45 t	***181.5/45 = 4.0 t CO_2eq/t product***	***181.5 t CO_2eq***

Mass-allocated PCF component	Emission factor	Emissions
Input A*, 25 t	1.0 t CO_2eq/t input A*	25 t CO_2eq
Heat, 45 GJ	0.1 t CO_2eq/t GJ heat	4.5 t CO_2eq
Emission, 11.3 CO_2	1.0 t CO_2eq/t CO_2	11.3 t CO_2eq
MB 100% Product, 22.5 t	***40.8/22.5 = 1.81***	***40.8 t CO_2eq***

Mass-allocated PCF component	Emission factor	Emissions
Input A, 75 t	3.0 t CO_2eq/t input A	225 t CO_2eq
Heat, 135 GJ	0.1 t CO_2eq/t GJ heat	13.5 t CO_2eq
Emission, 33.8 CO_2	1.0 t CO_2eq/t CO_2	33.8 t CO_2eq
Fossil product, 67.5 t	***272.3/67.5 = 4.0 t CO_2eq/t product***	***272.3 t CO_2eq***

Figure 9.
How do the "conventional reference" method and the "inventory" method differ in the calculation of PCF for mass-balance attributed products [47].

phase of the biomass feedstock. As a result, the net impact on atmospheric carbon levels can be considered neutral. This is, in contrast to fossil fuel-derived products, where the combustion releases carbon dioxide that was previously sequestered over millions of years underground.

11.1 The cradle-to-grave accounting approach

In a cradle-to-grave approach, which considers the entire life cycle of a product, from raw material sourcing to final product disposal, the biogenic carbon uptake[8] and emissions from disposal are balanced. This means that the CO_2 emissions associated with the end-of-life stage of biomass-derived products, such as composting or biodegradation, are offset by the biogenic carbon uptake during the growth phase of the bio-based renewable raw materials. Both carbon uptake and biogenic carbon dioxide emissions are considered in total as climate neutral.

This approach is often used for cradle-to-grave assessment of bio-based fuels. The specific characteristic of these product systems is that the combustion of fuel, and, therefore, the CO_2 release to the atmosphere, is inside the system boundaries. From this perspective, the biogenic CO_2 flows can be canceled out and the net carbon emissions from these products are zero or neutral (the so-called "0/0 accounting approach"). On the other hand, fossil-based products will receive a CO_2 burden in case of incineration or other forms of disposal. This is because the combustion of fossil-based products releases CO_2 that has been sequestered underground for millions of years, and if released now, contributing to the increase of CO_2 in the atmosphere.

11.2 The cradle-to-gate accounting approach

When a product's final disposal cannot be specifically identified, such as with platform chemicals used for various applications, or it is simply unknown, as is mainly the case for upstream suppliers in the value chain, a cradle-to-gate LCA is typically conducted. In a cradle-to-gate LCA, part of a product life cycle, i.e. the environmental impact of products from raw material extraction (cradle) to the point of leaving the factory gate (gate) are considered (i.e., before it is transported to the customer), **Figure 10**. Furthermore, this system boundary is selected in business-to-business relationships, when materials are delivered from suppliers to their customers. ISO 14067 [46] specifically focuses on carbon footprint assessment, including the treatment of biogenic carbon. In the context of PCF calculation, the −1/+1 approach for assimilated CO_2 should be applied, and the responsibilities and contributions of different actors along the value chains, such as biomass producers, processors, manufacturers, formulators, and end-users, made transparent. It is crucial to account for such scenarios where biogenic content may be retained or transferred to other materials or applications. **Figure 10** illustrates the different impacts throughout the product's life cycle until the factory gate.

The biogenic carbon remains in the molecule until it is released through incineration or degradation. For instance, in the case of indoor paint on a wall, the biogenic carbon will remain on the wall until the house is demolished or undergoes complete reconstruction. If the concrete of the house is repurposed for road construction or any other application, the biogenic content of the indoor paint may still be present in the material used for the initial purpose [53]. EN 15804 [54] is a European standard that specifically addresses the environmental impact assessment of construction

[8] Biogenic removals or biogenic uptake refers to the process of capturing and removing carbon dioxide (CO_2) from the atmosphere through the natural growth and absorption of carbon by plants, trees, and other living organisms. It is a mechanism by which living organisms, through photosynthesis, take in CO_2 and store it in their biomass or in the soil. Biogenic removals are considered a form of carbon sequestration and can help mitigate climate change by reducing the concentration of greenhouse gases in the atmosphere.

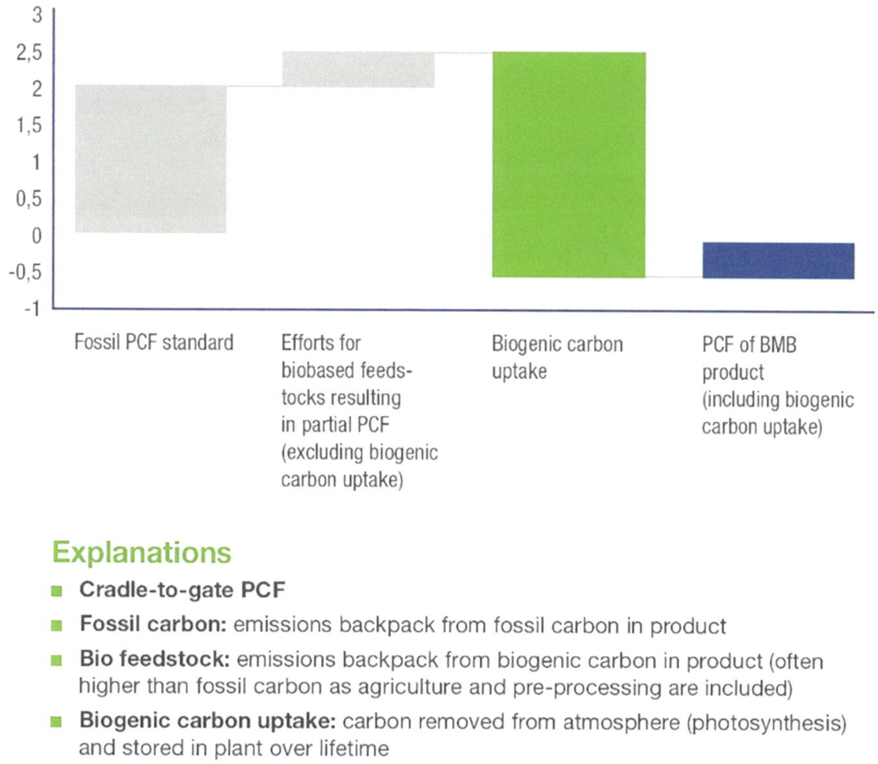

Figure 10.
Illustration of CO_2 contributions of a cradle-to-gate calculation of a bio-derived products.

products. The long-term benefits of biomass-derived products used in applications with extended lifespans or as investment goods need to be accounted for. Obligations for businesses under the CSRD [55] are to maintain transparency and traceability throughout the process, including data collection, reporting, and decision-making. Similarly, when biomass feedstocks are blended with fossil feedstocks, such as in mass balance scenarios, the greenhouse gas footprint difference between bio-attributed and fossil-based products needs to be captured [56].

In earlier versions of the Product Environmental Footprint (PEF) methodology [57], developed by the European Commission's Joint Research Centre (JRC) [58], the carbon removal potential of bio-derived products through photosynthesis and plant growth of biomass feedstock was not explicitly accounted for [59]. In **Figure 11**, the current methodology is illustrated. In **Figure 12**, the proposed alternative accounting method is illustrated. We describe a scenario, where there are two identical products with the only difference being the feedstock used (fossil-based and bio-based feedstock). The CO_2 footprint at the cradle will indeed differ. The bio-based feedstock, **Figure 12** will have a negative CO_2 emission (due to the uptake of CO_2 during biomass growth), while the fossil-based feedstock will have a zero CO_2 emission. Throughout the processing and use stages, the difference in the CO_2 emissions between the two feedstocks will become apparent. While the total carbon footprint at the end of life may be the same for both products, the bio-derived product will have a lower carbon footprint at any stage during processing and use, due to the negative CO_2 emission at the cradle.

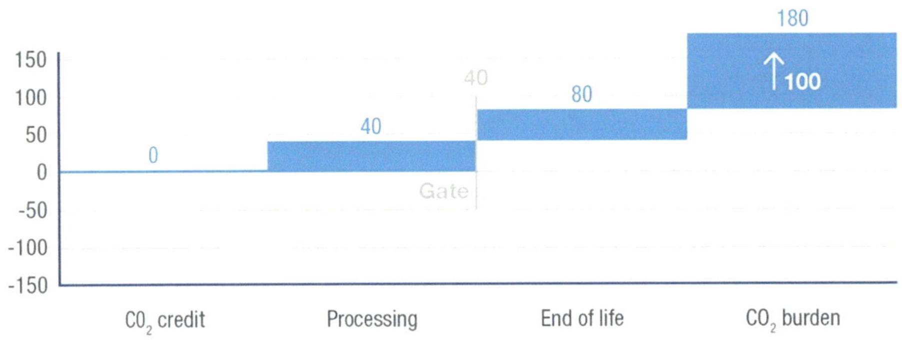

Figure 11.
Scenario where fossil-based feedstock is used, adopted from [60].

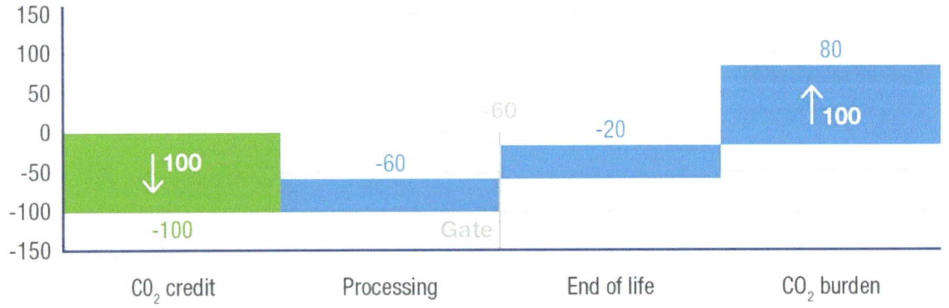

Figure 12.
Scenario where bio-based renewable raw material is used as feedstock, adopted from [60].

The omission of carbon sequestration in biomass-derived products was a limitation of earlier versions of the PEF methodology. The efforts to address limitations and improve the methodology's treatment of biogenic carbon are ongoing [61].

11.3 Driving change: The impact of policy in accelerating the transition to a sustainable future

Chemistry continues to be indispensable for modern life and the chemical industry successful transition requires clear steps, predictability, and instilling investors' confidence. The European Union has been actively promoting and supporting bioeconomy through its published strategies in 2012 and 2018 [62], with ongoing progress and support from sectorial and horizontal policies [63, 64]. The implementation of the principles of "cascading use" [65] of biomass is an important aspect for sustainable feedstock utilization. This approach emphasizes the efficient and strategic use of biomass, where the same biomass resource is used for multiple purposes, and in different value chains, maximizing its potential and reducing waste. Harmonized sustainability criteria are essential to ensure that the sourcing of biomass does not jeopardize local food production or compromise food security. In terms of carbon accounting, by applying the −1/+1 accounting, both on the foreground and on the background system level, the carbon sequestration potential via biomass, and the potential of the GHG emissions reduction can be recognized. Further details on general recommendations and recommendations by industry are expressed throughout the consultation process with association and sector groups [60, 66–71]. It is essential for stakeholders and practitioners to stay informed about the latest developments on methodologies like PEF and consult other relevant standards, where needed and appropriate. The ongoing update of the GHG Protocol [56], which provides guidelines for quantifying and reporting greenhouse gas emissions, should consider the need for methodological adaptation, mainly on the corporate reporting level [66].

12. Conclusion

The chemical industry serves as a crucial driver of innovation in a multitude of customer industries. The transformation of the chemical industry towards decarbonization requires a shift beyond the limitations of bio-based materials. The concept of bio-attributed materials offers a promising approach, leveraging existing infrastructure, to deliver affordable products without compromising performance. The CO_2 emissions and removals should be shown alongside the time periods that are in the scope of the calculations.

In summary, the adoption of cascading use principles, the consideration of sustainability criteria, the recognition and harmonization of biogenic carbon uptake accounting rule, the acknowledgement of the mass balance method, and the comprehensive product and corporate carbon dioxide accounting framework are all important elements in driving the transition towards environmentally sustainable production processes in the chemical industry.

Author details

Ivana Krkljuš*, Peter Saling and Christian Krüger
BASF SE, Ludwigshafen am Rhein, Germany

*Address all correspondence to: ivana.krkljus@basf.com

References

[1] Our Contribution to EU27 Industry. Available from: Our contribution to EU27 industry - cefic.org [Accessed: September 12, 2024]

[2] The European Green Deal: A Growth Strategy that Protects the Climate. Available from: The European Green Deal (europa.eu) [Accessed: September 12, 2024]

[3] 2023 Facts And Figures of the European Chemical Industry. Available from: 2023 Facts and Figures of the European Chemical Industry - cefic.org [Accessed: September 12, 2024]

[4] Facts and Figures of the European Chemical Industry. Available from: The-European-Chemical-Industry-Facts-And-Figures-2020.pdf (lademo.be) [Accessed: September 12, 2024]

[5] What the Draghi Report Says about EU Climate Action and Energy. Available from: Q&A: What the Draghi report says about EU climate action and energy - Carbon Brief [Accessed: September 12, 2024]

[6] The Chemical Industry Transition Pathway. Available from: EU Chemical Industry Transition Pathway (cefic.org) [Accessed: September 12, 2024]

[7] PlasticsEurope. Plastics—The Facts 2020. Available from: Plastics - the Facts 2020 • Plastics Europe [Accessed: September 12, 2024]

[8] Ellen MacArthur Foundation. The New Plastics Economy: Rethinking the Future of Plastics and Catalysing Action. Available from: The New Plastics Economy: Rethinking the future of plastics & catalysing action (ellenmacarthurfoundation.org) [Accessed: September 12, 2024]

[9] BP Statistical Review of World Energy 2020 | 69th Edition. Available from: Full report – BP Statistical Review of World Energy 2020 [Accessed: September 12, 2024]

[10] Eurostat. Recycling Rates for Plastic Packaging Waste. Available from: EU packaging waste generation with record increase - Eurostat (europa.eu) [Accessed: September 12, 2024]

[11] Geyer R, Jambeck JR, Law KL. Production, use, and fate of all plastics ever made. Science Advances. 2017;**3**:7. DOI: 10.1126/sciadv.1700782 [Accessed: November 20, 2024]

[12] RCI Carbon Flows Report – Compilation of Supply and Demand of Fossil and Renewable Carbon on a Global and European Level. Available from: RCI's scientific background report: "RCI carbon flows report – Compilation of supply and demand of fossil and renewable carbon on a global and European level" (Oct. 2023) | Renewable Carbon Publications (renewable-carbon.eu) [Accessed: September 12, 2024]

[13] World Steel Association. Steel Statistical Yearbook 2020. Available from: https://www.worldsteel.org/en/dam/jcr:177f3a3c-4f9b-4f0c-ae2d-30967b5cf6e7/Steel+Statistical+Yearbook+2020.pdf [Accessed: September 12, 2024]

[14] Bioeconomy in Europe: State of Play and Future Potential. European Commission. 2018. Available from: Bio-based economy in Europe - Publications Office of the EU (europa.eu) [Accessed: September 12, 2024]

[15] European Bioplastics. Global Bioplastics Market Report 2021. 2021. Available from:

Report_Bioplastics_Market_Data_2021_short_version.pdf (european-bioplastics.org) [Accessed: September 12, 2024]

[16] A Sustainable Bioeconomy for Europe: Strengthening the Connection between Economy, Society and the Environment: Updated Bioeconomy Strategy. Available from: A sustainable bioeconomy for Europe - Publications Office of the EU (europa.eu) [Accessed: September 12, 2024]

[17] What is Mass Balance and How Can it Help us Transition to a Circular Economy. Available from: How mass balance can support the circular economy | World Economic Forum (weforum.org) [Accessed: September 12, 2024]

[18] An Assessment of Mass Balance Accounting Methods for Polymers Workshop Report. Available from: An Assessment of Mass Balance Accounting Methods for Polymers Workshop Report | NIST [Accessed: September 12, 2024]

[19] Bioeconomy. Available from: Bioeconomy research and innovation (europa.eu) [Accessed: September 12, 2024]

[20] Ellen MacArthur Foundation, What is a Circular Economy? Available from: What is a circular economy? | Ellen MacArthur Foundation [Accessed: September 12, 2024]

[21] Ellen MacArthur Foundation. Why the World Needs a 'Circular Bioeconomy' - For Jobs, Biodiversity and Prosperity. Available from: What's a 'circular bioeconomy' and how can it save the planet? | World Economic Forum (weforum.org) [Accessed: September 12, 2024]

[22] ScienceDirect. Overview of Circular Bioeconomy. Available from: Circular Bioeconomy - An Overview | ScienceDirect Topics [Accessed: September 12, 2024]

[23] Available from: Circular Bioeconomy - An Overview | ScienceDirect Topics. Bio-Waste Management plays a Keyrole in Bioeconomy - European Compost Network [Accessed: September 12, 2024]

[24] ScienceDirect. Available from: Sustainable food waste management towards circular bioeconomy: Policy review, limitations and opportunities - ScienceDirect [Accessed: September 12, 2024]

[25] Springer, Circular Economy, Bioeconomy, and Sustainable Development Goals: A Systematic Literature Review. Available from: Circular economy, bioeconomy, and sustainable development goals: a systematic literature review | Environmental Science and Pollution Research (springer.com) [Accessed: September 12, 2024]

[26] ISO 14067:2018. Available from: ISO 14067:2018 - Greenhouse gases – Carbon footprint of products [Accessed: September 13, 2024]

[27] DIN EN 16575:2014-10. Available from: DIN EN 16575 - 2014-10 - DIN Media [Accessed: September 19, 2024]

[28] ASTM International, ASTM D6866-22. Available from: D6866 Standard Test Methods for Determining the Biobased Content of Solid, Liquid, and Gaseous Samples Using Radiocarbon Analysis (astm.org) [Accessed: September 13, 2024]

[29] DIN CEN/TS 16640:2014-05. Available from: DIN CEN/TS 16640 - 2014-05 - DIN Media [Accessed: September 13, 2024]

[30] DIN 51637:2014-02. Available from: DIN 51637 - 2014-02 - DIN Media [Accessed: September 13, 2024]

[31] Directive (EU) 2024/825 of the European Parliament and of the Council of 28 February 2024 amending Directives 2005/29/EC and 2011/83/EU as regards empowering consumers for the green transition through better protection against unfair practices and through better information. Available from: Directive - EU - 2024/825 - EN - EUR-Lex (europa.eu) [Accessed: September 13, 2024]

[32] EU Commission, Green Claims. Available from: Green claims - European Commission (europa.eu) [Accessed: September 13, 2024]

[33] ISO 22095:2020. Available from: ISO 22095:2020 - Chain of custody—General terminology and models [Accessed: Accessed: September 12, 2024]

[34] RECS association. Available from: https://recs.org/ [Accessed: September 13, 2024]

[35] DIN EN 16325 Guarantees of Origin related to energy - Guarantees of Origin for Electricity, gaseous hydrocarbons, Hydrogen, and heating and cooling; German and English version prEN 16325:2022. Available from: Drafts (din.de) [Accessed: September 17, 2024]

[36] Krüger C, Kicherer A, Kormann C, Raupp N. Biomass balance: An innovative and complementary method for using biomass as feedstock in the chemical industry. In: Benetto E, Gericke K, Guiton M, editors. Designing Sustainable Technologies, Products and Policies: From Science to Innovation. Cham: Springer International Publishing; 2018

[37] RedCert, Your Partner for Sustainability Certification. Available from: Scheme documents (redcert.org) [Accessed: August 08, 2024]

[38] ISCC System. Available from: ISCC System – Solutions for sustainable and deforestation free supply chains (iscc-system.org) [Accessed: September 13, 2024]

[39] ISO/CD 13662. Available from: https://www.iso.org/cms/%20render/live/en/sites/isoorg/contents/data/standard/08/44/84427.html [Accessed: September 13, 2024]

[40] Directive (EU) 2018/2001 of the European Parliament and of the Council of 11 December 2018 on the promotion of the use of energy from renewable sources (recast). Available from: EUR-Lex - 02018L2001-20240716 - EN - EUR-Lex (europa.eu) [Accessed: September 13, 2024]

[41] ISO 14040:2006 Environmental Management—Life Cycle Assessment—Principles and Framework. Available from: ISO 14040:2006 - Environmental management—Life cycle assessment—Principles and framework [Accessed: September 15, 2024]

[42] ISO 14044:2006 Environmental Management—Life Cycle Assessment—Requirements and Guidelines. Available from: ISO 14044:2006 - Environmental management—Life cycle assessment—Requirements and guidelines [Accessed: September 15, 2024]

[43] Jeswani H, Krüger C, Kicherer A, Anthony F, Azapagic A. A Methodology for Integrating the Biomass Balance Approach into Lifecycle Assessment with an Application in the Chemicals Sector. 2019. DOI: 10.1016/j.scitotenv.2019.06.088 [Accessed: September 15, 2024]

[44] LCA Summary Report for Eastman Methanolysis Technology (North America). Available from: PRT Methanolysis LCA Summary Report.pdf

(eastman.com) [Accessed: September 15, 2024]

[45] Life Cycle Assessment (LCA) for ChemCycling® and Measurement Program for Pyrolysis Oil. Available from: Life cycle assessment (LCA) for ChemCycling® (basf.com) [Accessed: September 15, 2024]

[46] ISO 14067:2018 Greenhouse Gases—Carbon Footprint of Products—Requirements and Guidelines for Quantification. Available from: ISO 14067:2018 - Greenhouse gases – Carbon footprint of products [Accessed: September 15, 2024]

[47] Together for Sustainability (TfS). Available from: Home - TFS Initiative (tfs-initiative.com) [Accessed: September 15, 2024]

[48] Together for Sustainability (TfS) PCF Guideline. Available from: TfS-PCF-Guidelines-2024.pdf [Accessed: December 20, 2024]

[49] Together for Sustainability (TfS) White Paper. Available from: TfS-2097-White-Paper-EN_hi_single_page_no_crop_.pdf (tfs-initiative.com) [Accessed: September 15, 2024]

[50] ISO 14020:2022 Environmental statements and programmes for products—Principles and general requirements. Available from: ISO 14020:2022—Environmental statements and programmes for products—Principles and general requirements [Accessed: September 15, 2024]

[51] EU Ecolabel. Available from: EU Ecolabel - Home (europa.eu) [Accessed: September 19, 2024]sssssssss

[52] ECOPlatform. Available from: Home - Eco Platform en (eco-platform.org) [Accessed: September 19, 2024]

[53] Saling P, Carey J, Assmann A, Engel A, Lohscheidt M, Scagnetti C, Isasa M. Indoor Paints for Concrete Walls Using Biomass Balance Binders: A Chemicals Case Study Applying the ORIENTING LCSA Methodology, European Union's Horizon 2020 Research and Innovation Programe under Grant Agreement No 958231, LCM Conference 2023, Lille. Available from: Provisional - Orienting [Accessed: September 17, 2024]

[54] DIN EN 15804:2022-03. Available from: DIN EN 15804 - 2022-03 - DIN Media [Accessed: September 15, 2024]

[55] Corporate Sustainability Reporting. Available from: Directive (EU) 2022/2464 of the European Parliament and of the Council of 14 December 2022 amending Regulation (EU) No 537/2014, Directive 2004/109/EC, Directive 2006/43/EC and Directive 2013/34/EU, as regards corporate sustainability reporting (Text with EEA relevance) [Accessed: September 15, 2024]

[56] GHG Protocol. Available from: Homepage | GHG Protocol [Accessed: September 15, 2024]

[57] Commission Recommendation (EU) 2021/2279 of 15 December 2021 on the use of the Environmental Footprint methods to measure and communicate the life cycle environmental performance of products and organisations. Available from: EUR-Lex - 32021H2279 - EN - EUR-Lex (europa.eu) [Accessed: September 15, 2024]

[58] Directorate-General JRC Joint Research Centre. Available from: Joint Research Centre (europa.eu) [Accessed: September 15, 2024]

[59] PANT Rana; CHOMKHAMSRI Kirana; PELLETIER Nathaniel; MANFREDI Simone; GALATOLA Michele; BEDO Imola, European Guidelines for measuring

the Environmental Footprint of Products and Organisations. Joint Research Centre, European Commission; 2012. Available at: JRC Publications Repository - European Guidelines for measuring the Environmental Footprint of Products and Organisations [Accessed: November 20, 2024]

[60] Cefic's counterproposal to assess the carbon footprint of bio-based products. Available from: Cefic's counterproposal to assess the carbon footprint of bio-based products - cefic.org [Accessed: August 08, 2024]

[61] European Commission Register of Environmental Footprint Technical Advisory Board (E03710) Meetings. Available from: Register of Commission expert groups and other similar entities (europa.eu) [Accessed: September 09, 2024]

[62] EU Commission Knowledge Center for Bioeconomy. Bioeconomy Strategy. Available from: | Knowledge for policy (europa.eu) [Accessed: September 09, 2024]

[63] Bioeconomy Globalization: Recent Trends and Drivers of National programs and Policies. Available from: Global_Bioeconomy_-_April_2024_IACGB.pdf [Accessed: August 08, 2024]

[64] Global Bioeconomy Assessment Coordinated Efforts of Policy, Innovation, and Sustainability for a Greener Future. Available from: 1713940443921.pdf (unep-iemp.org) [Accessed: August 08, 2024]

[65] European Commission Supporting Policy with Scientific Evidence. Available from: Cascading use | Knowledge for policy (europa.eu) [Accessed: September 15, 2024]

[66] Improving and Harmonizing Scope 3 Reporting TfS White Paper. Available from: TfS-2097-White-Paper-EN_hi_single_page_no_crop_.pdf (tfs-initiative.com) [Accessed: August 08, 2024]

[67] RCI's Scientific Background Report: "Non-Level Playing Field for Renewable Materials vs. Fossil in Life Cycle Assessments. Available from: https://renewable-carbon.eu/publications/product/rci-non-level-playing-field-for-renewable-materials-vs-fossil-in-lcas/ [Accessed: August 08, 2024]

[68] Joint Statement PEF TAB – Biogenic Carbon Modelling (PDF). Available from: Joint Statement PEF TAB – biogenic carbon modelling (PDF) | Renewable Carbon Publications (renewable-carbon.eu) [Accessed: August 08, 2024]

[69] Cefic Joint Statement. Available from: ttps://cefic.org/app/uploads/2024/09/Joint-Statement-on-biogenic-carbon-accounting-in-Product-Environmental-Footprint-PEF.pdf [Accessed: September 15, 2024]

[70] BioChem Europe Industry Position. Towards an Accurate Accounting for Carbon from Biomass in the Product Environmental Footprint (PEF). Available from: Towards an accurate accounting for carbon from biomass in the Product Environmental Footprint (PEF) - BioChem Europe (biochem-europe.eu) [Accessed: September 15, 2024]

[71] BioChem Europe Industry Position. Available from: https://wordpress.biochem-europe.eu/wp-content/uploads/2024/09/Industry-position-on-biogenic-carbon-accounting-in-PEF-final-version.pdf [Accessed: September 19, 2024]

Chapter 4

Biobased Activated Carbon and Its Application

Md Zakir Hossain and Muhammad Badrul Islam Chowdhury

Abstract

The increasing environmental concerns regarding the depletion of fossil resources and the excessive production of waste have shifted attention toward sustainable materials derived from renewable resources. Biobased activated carbon (BAC), derived from biomass, has emerged as a promising alternative to conventional fossil-derived activated carbon (AC), offering numerous advantages in terms of sustainability, cost-effectiveness, environmental impact, and wide-ranging applications. The production process of BAC involves the carbonization of biomass materials followed by activation, which enhances its porosity and surface area. These characteristics make BAC highly effective for applications in water and air purification, energy storage, and environmental remediation. In water treatment, BAC is used to remove pollutants like heavy metals, organic contaminants, and microplastics through adsorption. In air purification, it helps eliminate harmful gases and volatile organic compounds (VOCs). Additionally, BAC has emerged as a key material in energy storage technologies, particularly in supercapacitors, due to its high surface area and electrical conductivity. Its use in soil amendment and environmental remediation is also gaining attention for removing toxic substances from contaminated environments. The development of BAC is aligned with global efforts to reduce carbon footprints and promote circular economies. Its versatility and sustainability make BAC a promising material in addressing environmental challenges while providing an alternative to fossil fuel-derived products. This chapter will cover all the possible applications where BAC is being used.

Keywords: biomass, biobased activated carbon (BAC), renewable source, activation, emission

1. Introduction

Activated carbon (AC) is a porous material widely used in various industries, including water filtration and gas adsorption, due to its high surface area and exceptional adsorption properties [1]. AC has traditionally been derived from non-renewable sources such as coal, petroleum, and peat due to their high carbon content and availability. Coal, in particular, has been a primary source of activated carbon production [2]. However, with increasing environmental awareness and the depletion of fossil resources, attention has shifted toward biobased activated carbon (BAC).

Use of biomass has attracted considerable attention as an alternative resource because of its renewable and lasting nature, high utilization potential, abundance in nature, and low cost.

BAC can be produced from renewable biomass sources such as agricultural residues, wood, and waste materials [3, 4]. This is further supported by the idea of a 'zero waste economy,' which uses waste biomass as a feedstock to achieve this potential by converting it into value-added commodities [5, 6]. Converting these wastes into valuable products like activated carbon has the potential to solve environmental problems including agricultural waste residue, air and water pollution. Interestingly, biobased activated carbon often exhibits superior properties compared to coal-based activated carbon. For instance, biochar-activated carbon can achieve higher Brunauer–Emmett–Teller (BET) surface areas and iodine numbers [7]. Furthermore, substituting fossil fuels with activated carbon obtained from biomass will reduce the effects of global warming. The production of BAC has been shown to have significantly lower environmental impacts, with about 35% less cumulative energy demand and less than half the greenhouse gas emissions compared to coal-based activated carbon production [7, 8]. A life cycle assessment study comparing biomass-based and coal-based activated carbon production found that biomass-based activated carbon has a lower environmental impact in terms of global warming potential, acidification potential, and eutrophication potential [9].

The objective of this chapter is to provide a comprehensive summary of the applications of biobased activated carbon with a focus on the benefits of adopting more sustainable carbon sources. Its role in adsorption processes, environmental remediation, energy storage, and other industrial applications is highlighted to illustrate its multifaceted importance.

2. Biobased activated carbon

Activated carbon (AC) is a carbon-based material that has a wide variety of applications. Adsorption, purification, and catalytic uses of AC include food processing, medicines, beverages, metal recovery, and environmental appliances. It has been used to remove colorants and odors from municipal water sources, both underground and recycled. Common industrial applications with high AC consumption include the recovery of volatile organic compounds (VOCs), removal of toxic gases from flue gases, treatment of industrial wastewater, removal of siloxane from engine exhaust gases, recovery of precious metals and capture of landfill emissions. AC is mainly produced from two sources: coal and agricultural waste [10]. Based on the synthesis sources, AC is classified into two types: low-cost and commercial. Commercial AC is made from coal [11], apricot stone [12] and animal bones [13], whereas low-cost AC is made from renewable biomass resources such as date seed [14], rice husk [15], waste coffee ground [16], almond shell [17], corncob [18], waste tires [19], paulownia sawdust [20], sewage sludge [21], lotus stalk [22], peanut shell [23], coconut shell [24], longan shell [25], garlic peel [26], grass cuttings, horse manure, organic waste from beer production (beer waste) and bio-sludge [27], cane bagasse [28], hemp bast fibers [29], palm kernel shell [30], acorn shell [31], and Canarium Schwerin fruit nutshell [32]. Commercial activated carbon is more expensive than non-commercial activated carbon because it is produced from non-renewable raw materials [33]. As a result, research has been focused on preparing BAC from low-cost and environmentally friendly biomass sources, although the carbon content of these biomass precursors is lower than that of non-renewable precursors like coal, anthracite, or peat. Nonetheless, their availability, low cost, and non-harmful nature have a greater impact than their low carbon content and yields [34]. Several methods have been used to synthesize BAC from these sources such as chemical activation [23], physical (activation by steam and carbon dioxide) activation [14, 35], and hydrothermal carbonization [36]. A chemical

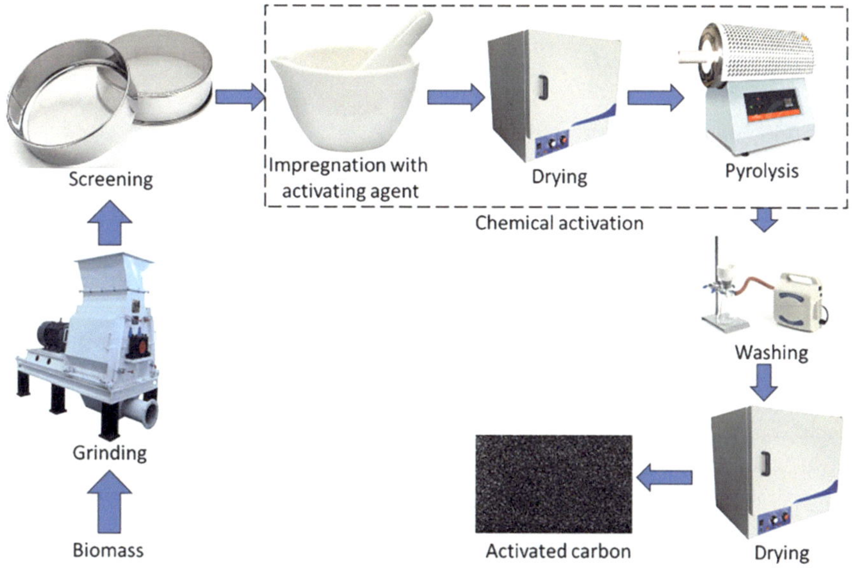

Figure 1.
Synthesis of BAC from biomass via chemical activation, redrawn from the reference [37].

Figure 2.
Synthesis of BAC from biomass via physical activation.

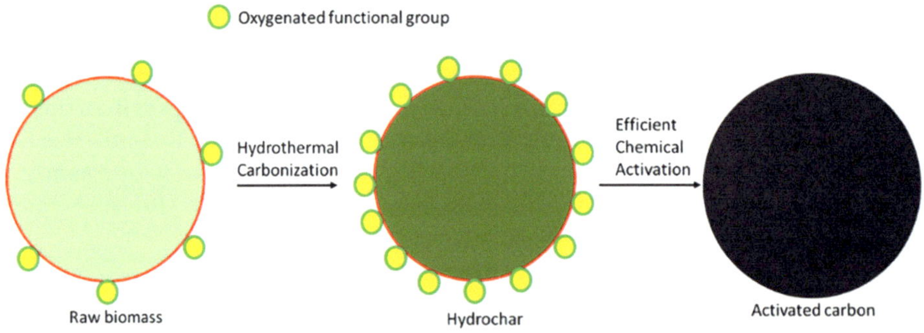

Figure 3.
Synthesis of BAC from biomass via hydrothermal carbonization, redrawn from the reference [36].

agent is impregnated with the raw/untreated biomass and then activated *via* pyrolysis at 500–600°C in the chemical activation process (**Figure 1**). Physical activation involves the carbonization of raw/untreated biomass at 800–1100°C temperature in the presence of steam or CO_2 (**Figure 2**). Hydrothermal carbonization uses wet biomass in an autoclave at a maximum temperature of 350°C for hydrochar production, followed by activation step using a chemical agent (**Figure 3**). Chemical activation is the most favorable method to prepare BAC compared to physical and hydrothermal carbonization methods. Large surface areas and higher AC yield were obtained using chemical activation method [38]. BAC is prepared through chemical activation using H_3PO_4 [39], H_2SO_4 [40], $ZnCl_2$ [41], $FeCl_3$ [42], NaOH, KOH [29], K_2CO_3 [43], CO_2 [44], steam [45], composite $ZnCl_2$–$CuCl_2$ [46], and H_2O–CO_2 [31].

3. Applications of biobased activated carbon

BAC offers a wide range of applications owing to its effective adsorption capabilities and environmental benefits. It is effective at removing contaminants from wastewater, including heavy metals (such as Pb and Cd), organic pollutants (such as pesticides and pharmaceuticals), and industrial wastes. This makes it an essential component of municipal and industrial water purification systems providing a long-term alternative to standard treatment methods.

BAC is utilized to capture and neutralize airborne pollutants, including volatile organic compounds (VOCs), odors, and harmful gases such as NH_3, H_2S, CO, and CO_2 to purify the air. Its ability to effectively adsorb these substances makes it valuable for maintaining indoor air quality and mitigating industrial emissions.

BAC is used in both residential and commercial water filters. It enhances the quality of drinking water by removing chlorine, sediment, and other impurities, thus ensuring safe and clean water for consumption. Its high surface area and porous structure also contribute to its effectiveness in treating well water and municipal supplies.

BAC is applied in processes such as decolorization and purification in the food and beverage industry. It helps in removing unwanted color, taste, and odor from products like sugar and alcohol, ensuring that the final product meets quality standards.

BAC is increasingly recognized for its potential in supercapacitor applications due to its excellent electrochemical properties and sustainability. BAC, derived from agricultural waste materials like coconut shells and wood, BAC features a highly porous structure with a large surface area, which enhances charge storage and transfer capabilities. In supercapacitors, it acts as an effective electrode material, providing high specific capacitance and energy density. The renewable nature of BAC, combined with its high performance, makes it a promising alternative to conventional carbon materials in energy storage devices, contributing to more sustainable and efficient power storage solutions.

The pharmaceutical industry utilizes BAC to purify drug compounds and remove contaminants during drug manufacturing processes. Its adsorption capabilities help in ensuring the purity and safety of pharmaceutical products.

Additionally, in the gold (Au) recovery industry, BAC is used in the extraction process to recover Au from ores and concentrates. Its effectiveness in adsorbing Au from cyanide solutions makes it a valuable material for improving recovery rates and reducing environmental impact.

In a nutshell, BAC offers a versatile and eco-friendly solution for various applications, providing effective performance while contributing to sustainability and waste reduction. This section explains the detailed applications of BAC.

3.1 Wastewater treatment

Water is one of the most critical natural resources on Earth, and it is essential for the survival of all living organisms. However, the growing global population, industrialization, and urbanization have led to increased pollution of water sources, making wastewater treatment an indispensable process. Wastewater treatment refers to the process of removing contaminants from sewage and industrial waste to make it safe for reuse or discharge into the environment. This process plays a pivotal role in public health protection, environmental sustainability, resource conservation, and economic efficiency.

BAC offers a sustainable and eco-friendly alternative for wastewater treatment by utilizing agricultural waste materials as feedstocks [47]. The production process involves pyrolysis, in which biomass is thermally decomposed in an oxygen-limited environment, and activation, which can be achieved using steam or chemical agents. This results in a highly porous structure with an expansive surface area and numerous micropores that are crucial for adsorption.

The unique properties of BAC allow it to effectively remove various contaminants from wastewater, including dyes [48], heavy metals (e.g., Pb, Cd, and Hg) [49, 50], organic pollutants (e.g., phenol, pesticides, pharmaceuticals, antibiotics, and industrial solvents) [51–53], and emerging pollutants that are difficult to degrade using traditional methods [54, 55]. **Tables 1–3** show the application of BAC obtained from different biomass sources which were used for dyes, metals, and organic pollutants removal from aqueous solution. Adsorption efficiency of BAC is often comparable to or even better than conventional activated carbon made from non-renewable sources like coal or wood due to the large surface area, tunable pore size distribution, and the presence of surface functional groups (such as -OH and -COOH groups), which enhance interaction with contaminants.

In addition to its adsorption performance, BAC is economically and environmentally advantageous. It repurposes agricultural byproducts, reduces waste and contributes to a circular economy, while also requiring lower production costs [87]. Furthermore, it can be regenerated and reused, maintaining high adsorption efficiency over multiple cycles, making it a viable option for long-term water purification

Biomass source	Target dye	References
Samanea saman waste pods	Rhodamine B	Kumbhar et al. [56]
Softwood bark	Orange 16 dye	Averheim et al. [48]
Sugarcane bagasse	Methylene blue	Sutthasupa et al. [57]
Walnut shell	Methylene blue	Vakili et al. [58]
Arachis hypogaea shell	Bromophenol blue	Raghunathan et al. [59]
Corncob	Methylene blue	Sun et al. [60]
Pomelo peel waste	Methylene blue	Zhang et al. [61]
Waste phenolic resin	Methylene blue	Hu et al. [62]

Table 1.
Dyes adsorption on BAC synthesized from different biomass sources.

Biomass source	Target metal	References
Samanea saman waste pods (SSWPAC)	Cr(VI)	Kumbhar et al. [56]
Melia azedarach leaves and twigs	Cu(II), Pb(II), and Ni(II)	Imran-Shaukat et al. [63]
Rice husk	Cu(II), Co(II), Zn(II), Pb(II), and Ni(II)	Olupot et al. [64]
Gas-to-liquids (GTL) derived biosolids, carboard and their mixed samples	Sr(II) and Ba(II)	Zuhara and McKay [65]
Brown alga	Pb(II)	Osman et al. [66]
Coconut shell	Zn(II) and K(I)	Packialakshmi et al. [47]
Commercial wood	Cr(VI)	Wu et al. [67]
Rice husk	Fe(III) and Mn(II)	Elewa et al. [68]
Lignocellulosic biomass	Cr(VI), Pb(II), Cu(II), Zn(II), Cd(II), Hg(III), As(V)	Hoang et al. [50]
Coconut shell	Zn(II)	Yusop et al. [69]
Forest and agricultural waste, industrial byproducts and waste, and municipal solid waste	Ag(I), Pb(II), Cu(II), Cd(II), As(V), Cr(VI), Ni(II), Hg(III)	Gupta et al. [70]
Bambusa vulgaris var. striata	Hg(III)	Mistar et al. [71]

Table 2.
Metals adsorption on BAC synthesized from different biomass sources.

systems. Given its potential to tackle both industrial and municipal wastewater issues, BAC represents a promising technology in the field of sustainable wastewater management.

3.2 Air treatment

BAC is increasingly being utilized in air purification systems due to its exceptional adsorption properties and environmental advantages. Made from renewable agricultural byproducts such as date stones [88], rambutan peel [89], palm kernel shell [90], fruit waste [91], mango seed shell [92] and pinewood sawdust [93], this type of BAC undergoes a rigorous activation process to develop a highly porous structure with a

Biomass source	Target compound	References
Pomegranate shell	Chlorpyrifos	Hussain et al. [72]
Erythrina speciosa	Paracetamol	Georgin et al. [73]
Antibiotic mycelial residues	Phenol	Wei et al. [74]
Residue of the fruit of Butiacapitate	Paracetamol Ketoprofenon	Yanan et al. [75]
Hazelnut	Imidacloprid	Mohammad and El-Refaey [76]
Red stems	Phenol	Zhang et al. [77]
Walnut	Imidacloprid	Mohammad and El-Refaey [76]
Acacia mangium	Phenol	Alam et al. [78]
Walnut shell	Imidacloprid	Motaghi et al. [79]
Peach stones	Imidacloprid	Mohammad and El-Sayed [80]
Baobab Fruit Shell	Phenol	Nedjai et al. [81]
Banyan tree roots	Phenol	Nirmala et al. [82]
Sewage sludge	Acetamiprid Thiamethoxam Imidacloprid	Sanz-Santos et al. [83]
Commercial coconut shell	Acetochlor	Wang et al. [84]
Rice husk	Phenol	Lv et al. [85]
Black wattle bark waste		Lütke et al. [86]

Table 3.
Organic pollutants (phenol, pesticides, pharmaceuticals) adsorption on BAC synthesized from different biomass sources.

vast surface area. This structural characteristic allows it to effectively capture a wide range of airborne contaminants, including volatile organic compounds (VOCs) [94], odors [95], and hazardous gases such as ammonia, hydrogen sulfide, carbon dioxide, and formaldehyde [96–102]. **Table 4** shows the application of BAC obtained from different biomass for VOCs and harmful gases adsorption. Its effectiveness in removing these pollutants helps improve indoor air quality and control industrial emissions, making it a valuable component in air filters, ventilation systems, and gas masks. Additionally, the use of biobased AC in air purification systems not only addresses pollution but also promotes sustainability by repurposing agricultural waste, reducing reliance on non-renewable resources, and minimizing environmental impact. Its high performance, combined with its eco-friendly benefits, positions biobased AC as a promising solution for achieving cleaner and healthier air.

3.3 Environmental remediation

Environmental remediation refers to the process of removing contaminants, pollutants, or hazardous materials from the soil, water, or air to restore ecosystems to a safe and sustainable state. It plays a critical role in addressing the negative impacts of industrial activities, oil spills, mining, and chemical waste disposal, which can severely harm biodiversity, human health, and the natural environment. Traditional remediation techniques, such as chemical treatments and synthetic adsorbents, often

Biomass source	Target compound	References
Seaweed waste	NH_3	Wang et al. [103]
Penicillin mycelial residues	Dichloromethane and chlorobenzene	Wei et al. [104]
Bamboo chip	Acetone, benzene, toluene, ethylbenzene, *ortho*-xylene, meta-xylene and para-xylene	Liu et al. [105]
Waste wood chip	Dichloromethane and toluene	Pi et al. [106]
Date stones	CO_2	Danish et al. [88]
Rambutan peel	CO_2	Zubbri et al. [89]
Palm kernel shell	CO_2	Rashidi et al. [90]
Fruit waste	CO_2	Serafin et al. [91]
Pinewood sawdust	CO_2	Quan et al. [93]
Banana peel	Benzene and toluene	Shen et al. [107]
Coconut shell	benzene, methanol, *n*-hexane and cyclohexane	Zhang et al. [108]
Hickory wood	Acetone and cyclohexane	Zhang et al. [109]

Table 4.
VOCs and harmful gases adsorption on BAC synthesized from different biomass sources.

exacerbate the environmental impact due to their non-biodegradability and high energy consumption. Thus, BAC presents a promising alternative for sustainable environmental remediation. BAC, known for its high surface area and porous structure, is widely used in adsorption processes to remove contaminants. BAC is particularly attractive because it is derived from renewable resources, such as agricultural waste, wood, and fruit peels. It provides a sustainable approach to waste valorization and environmental cleanup.

3.3.1 Water purification

Water contamination from industrial discharges, agriculture, and urban runoff is a major environmental challenge. BAC has proven effective in removing a wide range of pollutants, including the following.

Heavy metals: BAC can adsorb toxic metals such as lead (Pb), mercury (Hg), cadmium (Cd), and arsenic (As) through ion exchange and surface complexation (**Table 2**).

Organic pollutants: The porous structure of BAC allows it to adsorb dyes, pesticides, phenols, and pharmaceutical residues from wastewater (**Tables 1** and **3**).

Emerging contaminants removal: BAC has also shown potential efficiency for removing emerging contaminants (posing numerous public health concerns) from drinking water to deliver safe water. **Table 5** shows the potential of BAC for emerging pollutants adsorption from drinking water.

3.3.2 Air purification

BAC is widely used in air filtration systems to remove volatile organic compounds (VOCs) and other airborne pollutants. BAC has demonstrated efficacy in adsorption of VOCs (such as benzene and toluene), and odor control systems for industrial

Biomass source	Target compound	References
Almond shell	Bacteria (*E. coli*)	Belcheva et al. [110]
Rubber wood sawdust, ramie fiber, and areca husk	Bacteria (*E. coli*)	Nath et al. [111]
Raffia palm shells	Fluoride ion	Iwar et al. [112]
Bamboo	TSS, turbidity, and pH	Kuok et al. [113]
Coconut shell	Pathogens and disinfection byproducts	Xing et al. [114]
Almond, coconut, peanut, and walnut shells	Triazine (atrazine and prometon) and estriol	Xiao et al. [115]
Coconut shell	Dissolved organic carbon (DOC)	Chaukura et al. [116]
Coconut shell	BOD_5	Suif et al. [117]
Coconut shell	Total coliform count and *E. coli* bacteria	Yusuf and Murtala [118]
Coconut shell	Microcystins	Chaffin et al. [119]

Table 5.
Emerging contaminants adsorption on BAC synthesized from different biomass sources.

emissions as it adsorbs sulfur compounds, ammonia, and other odor-causing chemicals and explored for capturing carbon dioxide, contributing to efforts to mitigate climate change (**Table 4**).

3.3.3 Soil remediation

BAC can immobilize contaminants in soils, thereby reducing their bioavailability and leaching into groundwater. This method has applications in the remediation of heavy metals and organic pollutants in contaminated soils, particularly in areas near industrial sites and agricultural lands (**Tables 2** and **3**).

3.4 Food industry

The food industry is continuously evolving, driven by consumer demand for safe and high-quality products and the need for sustainable practices. Food safety concerns, including the presence of chemical contaminants, spoilage, and foodborne pathogens, pose a significant threat to public health. Traditional methods of food preservation and contamination control often rely on synthetic chemicals, which may have negative health and environmental impacts. BAC has emerged as a natural, sustainable alternative for various food-related applications. BAC is characterized by several properties that make it suitable for use in food applications:

High adsorption capacity: Due to its large surface area and porous structure, BAC can effectively adsorb contaminants, including chemical residues, toxins, and undesired flavors or odors.

Non-toxic and safe for consumption: BAC is considered safe for food use as it is produced from natural materials with no harmful chemicals involved in the production process.

Chemical stability: BAC remains stable in different food environments, whether in liquids, solids, or gases, making it versatile for various food processing applications.

BAC has been used for the following applications due to the presence of the above characteristics:

3.4.1 Food purification and decontamination

One of the most significant uses of BAC in the food industry is the purification of food products by removing contaminants such as:

Pesticide residues: BAC can adsorb harmful pesticide residues that remain on fruits, vegetables, and other agricultural products [120].

Chemical additives: In processed foods, BAC can help reduce residual chemicals from preservatives, flavor enhancers, and colorants [121].

Heavy metals: BAC is used in food and beverage processing to remove trace heavy metals such as Cadmium (Hg), lead (Pb), and cadmium (Cd), which can accumulate in food products and pose health risks [122, 123].

3.4.2 Decolorization and taste improvement

BAC plays a crucial role in improving the appearance and taste of food products:

Decolorization: BAC is widely used in sugar refining, where it adsorbs impurities and colorants, resulting in pure, white sugar [124, 125]. It is also applied in the production of fruit juices, wine, and edible oils to remove unwanted colors [126–128].

Odor and flavor control: BAC can be employed to remove off-flavors or odors in processed foods, making them more palatable. For example, in the production of beer, carbon filters can remove undesirable volatile compounds [129].

3.4.3 Food preservation

BAC contributes to extending the shelf life of perishable food products:

Adsorption of ethylene gas: BAC can adsorb ethylene gas, which is responsible for the ripening and aging of fruits and vegetables [130]. By controlling ethylene levels, BAC helps delay ripening and spoilage, thus extending shelf life.

Moisture and humidity control: In food packaging, BAC is used in desiccant packets to absorb excess moisture, thereby preventing spoilage and degradation of food products [131].

3.4.4 Food packaging

BAC plays an important role in advanced food packaging systems:

Active packaging: BAC is incorporated into packaging materials that actively absorb harmful gases, moisture, and odors, thereby improving the quality and safety of packaged foods [132].

Odor absorption: In meat, fish, and dairy packaging, BAC can be used to absorb unpleasant odors caused by the degradation of food compounds, thus enhancing consumer acceptance [133].

3.4.5 Water and beverage filtration

BAC is commonly used in the filtration of beverages and water for food production:

Water purification: Water used in food processing must be free from contaminants. BAC filters remove organic compounds, chlorine, and unwanted tastes or odors from water used in brewing, soft drink production, and food manufacturing [55].

Beverage filtration: Activated carbon is widely used in the beverage industry to purify and refine products such as beer, wine, spirits, and juices. It removes phenolic compounds, tannins, and other impurities that can affect the flavor, color, and clarity.

3.5 Pharmaceutical applications

Pharmaceutical compounds are the most frequently used chemicals in human, agricultural, and aquatic uses. Because of increased medication usage, aquatic ecosystems have been constantly exposed to pharmaceutical waste [134]. These chemicals can remain in water because of their extraordinary hydrophilicity and stability under aqueous conditions, which represents an environmental threat. Pharmaceuticals may still have a harmful impact on the environment, although their concentrations in water are generally negligible [135]. **Table 6** shows the application of BAC for the pharmaceutical compound's removal from the contaminated aqueous streams.

3.6 Energy storage applications

The transition to renewable energy sources has created a pressing need for efficient and sustainable energy storage technologies. Traditional energy storage systems, such as lithium-ion batteries and fuel cells, rely on materials that are often expensive, non-renewable, and environmentally detrimental. BAC has emerged as an eco-friendly, cost-effective alternative for energy storage applications.

BAC is widely recognized for its high surface area, tunable pore structure, and electrical conductivity, making it suitable for use in supercapacitors, batteries, and other energy storage devices. The ability to derive BAC not only contributes to energy storage efficiency but also promotes waste valorization, reducing the environmental impact. This section emphasizes the applications of BAC in the energy storage field, focusing on the potential of these materials to improve the energy storage efficiency and sustainability.

3.6.1 Supercapacitors

Supercapacitors, also known as electrochemical capacitors, are energy storage devices that store energy *via* electrostatic charge accumulation at the electrode-electrolyte interface. They offer a high-power density, rapid charge/discharge

Biomass source	Target pharmaceutical	References
Sunflower seed	Diclofenac Ibuprofen	Alvear-Daza et al. [136]
Agricultural waste	Norfloxacin	Bednárek et al. [137]
Banana peel	Amoxicillin Carbamazepine	Al-Sareji et al. [138]
Rubber fig tree leaves	Tetracycline	Vinayagam et al. [139]
Peony seeds shell	Ciprofloxacin	Liu et al. [140]
Sugar cane bagasse	Tetracycline	Xiong et al. [141]
Walnut shell	Sulfonamide	Geng et al. [142]
Seed shell	Naproxen	Mondal et al. [143]
Paper sludge	Enrofloxacin	Chowdhury et al. [144]

Table 6.
Pharmaceutical compounds adsorption on BAC synthesized from different biomass sources.

rates, and long cycle life. BAC is ideal for use in supercapacitors due to its high surface area, porous structure, and ability to efficiently adsorb electrolyte ions efficiently [145–147].

Electrochemical double layer capacitors (EDLCs): BAC is commonly used in EDLCs, where energy is stored in the electric double layer formed at the interface between the BAC electrode and the electrolyte. The high surface area and optimized pore size distribution of BAC enhance the energy storage capacity of EDLCs [148, 149].

Hybrid supercapacitors: In hybrid supercapacitors, BAC can be combined with other materials, such as metal oxides or conducting polymers, to improve energy storage performance. These hybrid systems take advantage of the high-power density of supercapacitors and the high energy density of battery-like materials [150, 151].

3.6.2 Lithium-ion batteries (LIBs)

Lithium-ion batteries are widely used in portable electronics, electric vehicles, and renewable energy storage systems. The performance of LIBs largely depends on the properties of the electrode materials. BAC has been investigated as both anode and cathode materials in LIBs due to its high surface area, good conductivity, and ability to form stable interfaces with lithium ions [152, 153].

Anode materials: In LIBs, BAC can serve as an anode material, providing a large surface area for lithium-ion intercalation. The porous structure of BAC allows for rapid ion diffusion and minimizes the risk of volume expansion, which is a common issue in traditional anode materials [154, 155].

Cathode materials: While activated carbon is more commonly used as an anode material, it has also been explored as a component in composite cathodes, where it can improve electron conductivity and enhance the overall electrochemical performance of the battery [156].

3.6.3 Sodium-ion batteries (SIBs)

Sodium-ion batteries are considered a promising alternative to lithium-ion batteries due to the abundance and low cost of sodium compared to lithium. However, the larger ionic radius of sodium poses challenges in terms of electrode materials. BAC, with its highly porous structure and tunable pore sizes, can accommodate sodium ions effectively, making it a potential candidate for SIBs [157].

Anode materials: Biomass-based activated carbon can be used as an anode material in SIBs, providing a stable structure for sodium-ion intercalation and deintercalation. The surface functional groups and large surface area of the activated carbon enhance the electrochemical performance of the anode [158, 159].

Composite electrodes: In sodium-ion batteries, biomass-based activated carbon can be combined with other materials, such as carbon nanotubes or graphene, to form composite electrodes with improved conductivity and energy storage capacity [160, 161].

3.6.4 Fuel cells

Fuel cells convert chemical energy directly into electrical energy, making them highly efficient energy conversion devices. BAC can be used as a catalyst support material in fuel cells due to its high surface area, porosity, and ability to disperse catalytic nanoparticles effectively [162].

Proton exchange membrane fuel cells (PEMFCs): In PEMFCs, biomass-based activated carbon can serve as a support material for platinum-based catalysts, enhancing the dispersion of catalyst particles and improving the overall efficiency of the fuel cell [163].

Direct methanol fuel cells (DMFCs): Activated carbon from biomass can also be used in DMFCs, where it serves as a support material for electrocatalysts, promoting the oxidation of methanol and improving the performance of the fuel cell [164, 165].

4. Conclusion

In conclusion, BAC has emerged as a sustainable and cost-effective alternative to conventional activated carbon derived from non-renewable resources. Utilizing biomass waste materials such as agricultural byproducts, forestry residues, and industrial biomass, this approach not only promotes environmental sustainability but also adds value to what would otherwise be discarded as waste.

The versatility of BAC in various applications, including water purification, air filtration, energy storage (e.g., supercapacitors), and catalysis, demonstrates its effectiveness due to its high surface area, porosity, and customizable chemical properties. In environmental remediation, it plays a crucial role in adsorbing contaminants such as heavy metals, organic pollutants, and gases. The production of BAC also supports the global movement toward circular economies by enhancing resource recovery and reducing the carbon footprint associated with traditional activated carbon production.

However, challenges remain in scaling up production and reducing energy consumption, warranting further research and innovation in this area. Future research and development in optimizing production techniques, enhancing adsorption capacities, and expanding its range of applications will further solidify BAC as a vital material in addressing global environmental and energy challenges.

Acknowledgements

The corresponding author gratefully acknowledges the fellowship awarded by MITACS, Canada.

Declaration of generative AI in scientific writing

During the preparation of this work, the author(s) used ChatGPT (version 3) to improve the quality of writing, sentence structure and vocabulary. After using this tool/service, the author(s) reviewed and edited the contents as needed and takes full responsibility for the content of the publication.

Author details

Md. Zakir Hossain[1*] and Muhammad Badrul Islam Chowdhury[2]

1 Department of Chemical and Biochemical Engineering, Western University, London, ON, Canada

2 CanmetENERGY Devon, Natural Resources of Canada, Devon, Canada

*Address all correspondence to: mhossa32@uwo.ca

References

[1] Churchill JG, Mumford KA. Granular activated carbon in water treatment. In: Pure and Functionalized Carbon Based Nanomaterials. Boca Raton: CRC Press; 2020. pp. 298-325

[2] Simate GS et al. Coal-based adsorbents for water and wastewater treatment. Journal of Environmental Chemical Engineering. 2016;**4**(2):2291-2312

[3] Raghu Vamshi Krishna B, Rao TN, Murthy HA. Orange peel-derived activated carbon as an electrode material for supercapacitor application using nickel mesh. Chemical Papers. 2023;**77**(9):4953-4962

[4] Colomba A, Berruti F, Briens C. Model for the physical activation of biochar to activated carbon. Journal of Analytical and Applied Pyrolysis. 2022;**168**:105769

[5] Rajak RC, Jacob S, Kim BS. A holistic zero waste biorefinery approach for macroalgal biomass utilization: A review. Science of the Total Environment. 2020;**716**:137067

[6] De Bhowmick G, Sarmah AK, Sen R. Zero-waste algal biorefinery for bioenergy and biochar: A green leap towards achieving energy and environmental sustainability. Science of the Total Environment. 2019;**650**:2467-2482

[7] Gu H et al. Life-cycle assessment of activated carbon from woody biomass. Wood and Fiber Science. 2018;**50**(3):229-243

[8] Shaheen J, Fseha YH, Sizirici B. Performance, life cycle assessment, and economic comparison between date palm waste biochar and activated carbon derived from woody biomass. Heliyon. 2022;**8**(12):e12388

[9] Yong JB, Tan LS, Tan J. Comparative life cycle assessment of biomass-based and coal-based activated carbon production. Progress in Energy and Environment. 2022:1-15

[10] Ukanwa KS et al. A review of chemicals to produce activated carbon from agricultural waste biomass. Sustainability. 2019;**11**(22):6204

[11] Zhao H et al. Coal based activated carbon nanofibers prepared by electrospinning. Journal of Materials Chemistry A. 2014;**2**(24):9338-9344

[12] Abbas M, Kaddour S, Trari M. Kinetic and equilibrium studies of cobalt adsorption on apricot stone activated carbon. Journal of Industrial and Engineering Chemistry. 2014;**20**(3):745-751

[13] Prabu D et al. Feasibility of magnetic nano adsorbent impregnated with activated carbon from animal bone waste: Application for the chromium (VI) removal. Environmental Research. 2022;**203**:111813

[14] Ogungbenro AE et al. Physical synthesis and characterization of activated carbon from date seeds for CO2 capture. Journal of Environmental Chemical Engineering. 2018;**6**(4):4245-4252

[15] Boonpoke A et al. Synthesis of activated carbon and MCM-41 from bagasse and rice husk and their carbon dioxide adsorption capacity. The Journal of Sustainable Energy and Environment. 2011;**2**(2):77-81

[16] Travis W, Gadipelli S, Guo Z. Superior CO 2 adsorption from waste coffee ground derived carbons. RSC Advances. 2015;**5**(37):29558-29562

[17] McCaffrey Z et al. Almond and walnut shell activated carbon for methylene blue adsorption. ACS Sustainable Resource Management. 2024;**1**(7):1421-1431

[18] Huarote-Garcia E et al. Activated carbon electrodes for supercapacitors from purple corncob (*Zea mays* L.). ACS Environmental Au. 2024;**4**(2):80-88

[19] Ahmad W et al. Adsorption of hydrocarbon pollutants from wastewater using Cu-and Zn-loaded activated carbon derived from waste tires. Environmental Progress and Sustainable Energy. 2024;**43**(3):e14360

[20] Zhu X-L et al. Activated carbon produced from paulownia sawdust for high-performance CO2 sorbents. Chinese Chemical Letters. 2014;**25**(6):929-932

[21] Martínez-Alvarenga H et al. Integral evaluation of effective conversion of sewage sludge from WWTP into highly porous activated carbon. Journal of Environmental Management. 2024;**351**:119822

[22] Liu H et al. Preparation of activated carbon from lotus stalks with the mixture of phosphoric acid and pentaerythritol impregnation and its application for Ni (II) sorption. Chemical Engineering Journal. 2012;**209**:155-162

[23] Fletcher A, Somorin T, Aladeokin O. Production of high surface area activated carbon from peanut shell by chemical activation with zinc chloride: Optimisation and characterization. Bioenergy Research. 2024;**17**(1):467-478

[24] Singla MK et al. Characterization of an Activated Carbon Electrode Made from Coconut Shell Precursor for Hydrogen Storage Applications. International Journal of Hydrogen Energy. 2024;**3**(61):1417-1428

[25] Wei H et al. Excellent electrochemical properties and large CO2 capture of nitrogen-doped activated porous carbon synthesised from waste longan shells. Electrochimica Acta. 2017;**231**:403-411

[26] Huang G-G et al. Activated carbons prepared by the KOH activation of a hydrochar from garlic peel and their CO2 adsorption performance. New Carbon Materials. 2019;**34**(3):247-257

[27] Hao W et al. Activated carbons prepared from hydrothermally carbonized waste biomass used as adsorbents for CO2. Applied Energy. 2013;**112**:526-532

[28] Molza A et al. Biomass-derived carbons as friction reducing additives for lubricants: Tribological properties of biochars and activated carbons obtained from sugar cane bagasse. Lubricants. 2024;**12**(9):308

[29] Hossain MZ et al. High-surface-area mesoporous activated carbon from hemp bast fiber using hydrothermal processing. C. 2018;**4**(3):38

[30] Md, Zaini MS, Al-Junid SAM, Syed-Hassan SSA. Effects of nitrogen doping on lithium polysulfide anchoring by activated carbon derived from palm kernel shell. Journal of Porous Materials. 2024;**31**:1-12

[31] Şahin Ö, Saka C. Preparation and characterization of activated carbon from acorn shell by physical activation with H2O–CO2 in two-step pretreatment. Bioresource Technology. 2013;**136**:163-168

[32] Olawale AS et al. Preparation of phosphoric acid activated carbons from *Canarium Schweinfurthii* nutshell and its role in methylene blue adsorption. Journal of Chemical Engineering and Material Science. 2015;**6**(2):9-14

[33] Chen Y et al. Application studies of activated carbon derived from rice husks produced by chemical-thermal process—A review. Advances in Colloid and Interface Science. 2011;**163**(1):39-52

[34] Yahya MA, Al-Qodah Z, Ngah CZ. Agricultural bio-waste materials as potential sustainable precursors used for activated carbon production: A review. Renewable and Sustainable Energy Reviews. 2015;**46**:218-235

[35] Ullah S et al. Activated carbon derived from biomass for wastewater treatment: Synthesis, application and future challenges. Journal of Analytical and Applied Pyrolysis. 2024;**179**:106480

[36] Jain A, Balasubramanian R, Srinivasan M. Hydrothermal conversion of biomass waste to activated carbon with high porosity: A review. Chemical Engineering Journal. 2016;**283**:789-805

[37] Kumari G, Soni B, Karmee SK. Synthesis of activated carbon from groundnut shell via chemical activation. Journal of The Institution of Engineers (India): Series E. 2022;**103**(1):15-22

[38] Neme I, Gonfa G, Masi C. Activated carbon from biomass precursors using phosphoric acid: A review. Heliyon. 2022;**8**(12):e11940

[39] Luo Y et al. The performance of phosphoric acid in the preparation of activated carbon-containing phosphorus species from rice husk residue. Journal of Materials Science. 2019;**54**(6):5008-5021

[40] Jawad AH et al. Adsorption of methylene blue onto activated carbon developed from biomass waste by H2SO4 activation: Kinetic, equilibrium and thermodynamic studies. Desalination and Water Treatment. 2016;**57**(52):25194-25206

[41] Gurung M et al. Hemp fibers derived porous carbon for naphthenic acids removal from contaminated aqueous stream. Biomass Conversion and Biorefinery. 2024:1-15

[42] Jawaduddin M et al. Synthetic grey water treatment through FeCl3-activated carbon obtained from cotton stalks and river sand. Civil Engineering Journal. 2019;**5**(2):340-348

[43] Abdulhameed AS et al. Statistical modeling and mechanistic pathway for methylene blue dye removal by high surface area and mesoporous grass-based activated carbon using K2CO3 activator. Journal of Environmental Chemical Engineering. 2021;**9**(4):105530

[44] Pallarés J, González-Cencerrado A, Arauzo I. Production and characterization of activated carbon from barley straw by physical activation with carbon dioxide and steam. Biomass and Bioenergy. 2018;**115**:64-73

[45] Zhou J, Luo A, Zhao Y. Preparation and characterisation of activated carbon from waste tea by physical activation using steam. Journal of the Air and Waste Management Association. 2018;**68**(12):1269-1277

[46] Bian Y et al. Recycling of waste sludge: Preparation and application of sludge-based activated carbon. International Journal of Polymer Science. 2018;**2018**(1):8320609

[47] Packialakshmi S et al. Treatment of industrial wastewater using coconut shell

based activated carbon. Materials Today: Proceedings. 2023;**81**:1167-1171

[48] Averheim A et al. Enhanced biobased carbon materials made from softwood bark via a steam explosion preprocessing step for reactive orange 16 dye adsorption. Bioresource Technology. 2024;**400**:130698

[49] Gupta VK et al. A critical analysis on the efficiency of activated carbons from low-cost precursors for heavy metals remediation. Critical Reviews in Environmental Science and Technology. 2015;**45**(6):613-668

[50] Hoang AT et al. Remediation of heavy metal polluted waters using activated carbon from lignocellulosic biomass: An update of recent trends. Chemosphere. 2022;**302**:134825

[51] Wang Y-H et al. Activated carbon fibers via reductive carbonization of cellulosic biomass for adsorption of nonpolar volatile organic compounds. Colloids and Surfaces A: Physicochemical and Engineering Aspects. 2021;**612**:125908

[52] Shamsollahi Z, Partovinia A. Recent advances on pollutants removal by rice husk as a bio-based adsorbent: A critical review. Journal of Environmental Management. 2019;**246**:314-323

[53] Lewoyehu M. Comprehensive review on synthesis and application of activated carbon from agricultural residues for the remediation of venomous pollutants in wastewater. Journal of Analytical and Applied Pyrolysis. 2021;**159**:105279

[54] Li W et al. Efficient ex-situ catalytic upgrading of biomass pyrolysis vapors to produce methylfurans and phenol over bio-based activated carbon. Biomass and Bioenergy. 2020;**142**:105794

[55] Rathi BS, Renuka V. Recent research progress on modified activated carbon from biomass for the treatment of wastewater: A critical review. Environmental Quality Management. 2024;**33**(4):907-928

[56] Kumbhar P et al. Biobased carbon for effective removal of rhodamine B and Cr (VI) from aqueous solution: Kinetic, isotherm and thermodynamic study. Biomass Conversion and Biorefinery. 2024;**14**(3):3535-3550

[57] Sutthasupa S et al. Sugarcane bagasse-derived granular activated carbon hybridized with ash in bio-based alginate/gelatin polymer matrix for methylene blue adsorption. International Journal of Biological Macromolecules. 2023;**253**:127464

[58] Vakili A et al. The impact of activation temperature and time on the characteristics and performance of agricultural waste-based activated carbons for removing dye and residual COD from wastewater. Journal of Cleaner Production. 2023;**382**:134899

[59] Raghunathan G, Rajagopal P, Chithra K. Arachis hypogaea shell–an approach towards waste utilization for cationic and anionic dye removal. Cleaner Engineering and Technology. 2022;**6**:100385

[60] Sun Z et al. Corncob-derived activated carbon for efficient adsorption dye in sewage. ES Food and Agroforestry. 2021;**4**:61-74

[61] Zhang B, Wu Y, Cha L. Removal of methyl orange dye using activated biochar derived from pomelo peel wastes: Performance, isotherm, and kinetic studies. Journal of Dispersion Science and Technology. 2020;**41**(1):125-136

[62] Hu W et al. Waste phenolic resin derived activated carbon by

microwave-assisted KOH activation and application to dye wastewater treatment. Green Processing and Synthesis. 2019;**8**(1):408-415

[63] Imran-Shaukat M, Wahi R, Ngaini Z. Bio-based activated carbon for Cu (II), Ni (II), and Pb (II) ions aqueous sequestration: Physiochemical characterization, isotherms, and kinetics. International journal of Environmental Science and Technology. 2024:1-24

[64] Olupot PW et al. Optimization of heavy metal removal by activated carbon obtained as a co-product from fast pyrolysis of rice husks. Results in Materials. 2024;**21**:100545

[65] Zuhara S, McKay G. Waste-derived activated carbons for effective adsorptive removal of strontium, barium, and binary pollutants: A response surface methodology study. Journal of Environmental Chemical Engineering. 2024;**12**(3):112836

[66] Osman NS et al. Bio-removal of lead (II) ions under optimal condition by zinc chloride-impregnated activated carbon from brown alga. Biomass Conversion and Biorefinery. 2024:1-11

[67] Wu Z et al. Synthesis of novel magnetic activated carbon for effective Cr (VI) removal via synergistic adsorption and chemical reduction. Environmental Technology and Innovation. 2023;**30**:103092

[68] Elewa AM et al. Chemically activated carbon based on biomass for adsorption of Fe (III) and Mn (II) ions from aqueous solution. Materials. 2023;**16**(3):1251

[69] Yusop MFM, Jaya EMJ, Ahmad MA. Single-stage microwave assisted coconut shell based activated carbon for removal of Zn (II) ions from aqueous solution–optimization and batch studies. Arabian Journal of Chemistry. 2022;**15**(8):104011

[70] Gupta S et al. Latest trends in heavy metal removal from wastewater by biochar based sorbents. Journal of Water Process Engineering. 2020;**38**:101561

[71] Mistar EM et al. Adsorption of mercury (II) using activated carbon produced from Bambusa vulgaris var. striata in a fixed-bed column. Sains Malaysiana. 2019;**48**(4):719-725

[72] Hussain OA et al. Preparation and characterization of activated carbon from agricultural wastes and their ability to remove chlorpyrifos from water. Toxicology Reports. 2023;**10**:146-154

[73] Georgin J et al. Adsorption of the first-line covid treatment analgesic onto activated carbon from residual pods of *Erythrina S peciosa*. Environmental Management. 2023;**71**(4):795-808

[74] Wei X et al. Adsorption of phenol from aqueous solution on activated carbons prepared from antibiotic mycelial residues and traditional biomass. Fuel Processing Technology. 2023;**242**:107663

[75] Yanan C et al. Adsorption of paracetamol and ketoprofenon activated charcoal prepared from the residue of the fruit of Butiacapitate: Experiments and theoretical interpretations. Chemical Engineering Journal. 2023;**454**:139943

[76] Mohammad SG, El-Refaey AA. Efficient removal of imidacloprid pesticide by two eco-friendly activated carbons. Journal of Water And Land Development. 2023;**57**

[77] Zhang X et al. Magnetic reed biochar materials as adsorbents for aqueous copper and phenol removal.

Environmental Science and Pollution Research. 2023;**30**(2):3659-3667

[78] Alam MG et al. Response surface methodology approach of phenol removal study using high-quality activated carbon derived from H3PO4 activation of Acacia mangium wood. Diamond and Related Materials. 2023;**132**:109632

[79] Motaghi H et al. Simultaneous adsorption of cobalt ions, azo dye, and imidacloprid pesticide on the magnetic chitosan/activated carbon@ UiO-66 bio-nanocomposite: Optimization, mechanisms, regeneration, and application. Separation and Purification Technology. 2022;**284**:120258

[80] Mohammad SG, El-Sayed MM. Removal of imidacloprid pesticide using nanoporous activated carbons produced via pyrolysis of peach stone agricultural wastes. Chemical Engineering Communications. 2021;**208**(8):1069-1080

[81] Nedjai R, Kabbashi NA, Alam MZ. Removal of phenol from aqueous solution by adsorption onto baobab fruit shell activated carbon: Equilibrium and kinetics studies. Journal of Environmental Treatment Techniques. 2021;**9**(3):686-697

[82] Nirmala G et al. Adsorptive removal of phenol using banyan root activated carbon. Chemical Engineering Communications. 2021;**208**(6):831-842

[83] Sanz-Santos E et al. Application of sludge-based activated carbons for the effective adsorption of neonicotinoid pesticides. Applied Sciences. 2021;**11**(7):3087

[84] Wang Y et al. Efficient removal of acetochlor pesticide from water using magnetic activated carbon: Adsorption performance, mechanism, and regeneration exploration. Science of the Total Environment. 2021;**778**:146353

[85] Lv S et al. A functional activated carbon for efficient adsorption of phenol derived from pyrolysis of rice husk, KOH-activation and EDTA-4Na-modification. Applied Surface Science. 2020;**510**:145425

[86] Lütke SF et al. Preparation of activated carbon from black wattle bark waste and its application for phenol adsorption. Journal of Environmental Chemical Engineering. 2019;**7**(5):103396

[87] Tibor ST, Grande CA. Industrial production of activated carbon using circular bioeconomy principles: Case study from a Romanian company. Cleaner Engineering and Technology. 2022;**7**:100443

[88] Danish M, Parthasarthy V, Al Mesfer MK. CO 2 capture using activated carbon synthesized from date stone: Breakthrough, equilibrium, and mass-transfer zone. Carbon Letters. 2021;**31**:1-12

[89] Zubbri NA et al. Low temperature CO2 capture on biomass-derived KOH-activated hydrochar established through hydrothermal carbonization with water-soaking pre-treatment. Journal of Environmental Chemical Engineering. 2021;**9**(2):105074

[90] Rashidi NA, Bokhari A, Yusup S. Evaluation of kinetics and mechanism properties of CO 2 adsorption onto the palm kernel shell activated carbon. Environmental Science and Pollution Research. 2021;**28**:33967-33979

[91] Serafin J et al. Conversion of fruit waste-derived biomass to highly microporous activated carbon

for enhanced CO2 capture. Waste Management. 2021;**136**:273-282

[92] Correia LB et al. CO 2 capture on activated carbons derived from mango fruit (*Mangifera indica L.*) seed shells: A TG study. Journal of Thermal Analysis and Calorimetry. 2018;**131**:579-586

[93] Quan C, Su R, Gao N. Preparation of activated biomass carbon from pine sawdust for supercapacitor and CO2 capture. International Journal of Energy Research. 2020;**44**(6):4335-4351

[94] David E. Production of activated biochar derived from residual biomass for adsorption of volatile organic compounds. Materials. 2022;**16**(1):389

[95] Wang S et al. Facile ultrasonic-assisted one-step preparation of sugarcane bagasse carbon sorbent for bio-based odor removal cat litter formulation. Industrial Crops and Products. 2022;**187**:115493

[96] Aghel B, Behaein S, Alobaid F. CO2 capture from biogas by biomass-based adsorbents: A review. Fuel. 2022;**328**:125276

[97] Phalakornkule C, Foungchuen J, Pitakchon T. Impregnation of chitosan onto activated carbon for high adsorption selectivity towards CO2: CO2 capture from biohydrogen, biogas and flue gas. Journal of Sustainable Energy and Environment. 2012;**3**(4):153-157

[98] Sawalha H et al. Removal of hydrogen sulfide from biogas using activated carbon synthesized from different locally available biomass wastes-a case study from Palestine. Bioengineered. 2020;**11**(1):607-618

[99] Tuerhong T, Kuerban Z. Preparation and characterization of cattle manure-based activated carbon for hydrogen sulfide removal at room temperature. Journal of Environmental Chemical Engineering. 2022;**10**(2):107177

[100] Abuelnoor N et al. Activated carbons from biomass-based sources for CO2 capture applications. Chemosphere. 2021;**282**:131111

[101] Bai W et al. Rice husk-based adsorbents for removing ammonia: Kinetics, thermodynamics and adsorption mechanism. Journal of Environmental Chemical Engineering. 2021;**9**(4):105793

[102] Isinkaralar K, Gullu G, Turkyilmaz A. Experimental study of formaldehyde and BTEX adsorption onto activated carbon from lignocellulosic biomass. Biomass Conversion and Biorefinery. 2023;**13**(5):4279-4289

[103] Wang S et al. Ammonia gas adsorption study using copper impregnated on mesoporous activated carbon from seaweed waste for indoor air purification. Building and Environment. 2024;**261**:111737

[104] Wei X et al. Preparation of N-doped activated carbons from penicillin mycelial residues and melamine for Cl-VOCs adsorption under humidity conditions. Journal of Environmental Chemical Engineering. 2023;**11**(5):110809

[105] Liu X et al. Role of molecular size of volatile organic compounds on their adsorption by KOH-activated micro-mesoporous carbon. Journal of Hazardous Materials. 2022;**424**:127355

[106] Pi X et al. Catalytic activation preparation of nitrogen-doped hierarchical porous bio-char for efficient adsorption of dichloromethane and toluene. Journal of Analytical and Applied Pyrolysis. 2021;**156**:105150

[107] Shen X et al. Record-high capture of volatile benzene and toluene enabled by activator implant-optimized banana peel-derived engineering carbonaceous adsorbents. Environment International. 2020;**143**:105774

[108] Zhang Z et al. Micro-mesoporous activated carbon simultaneously possessing large surface area and ultra-high pore volume for efficiently adsorbing various VOCs. Carbon. 2020;**170**:567-579

[109] Zhang X et al. Chemically activated hydrochar as an effective adsorbent for volatile organic compounds (VOCs). Chemosphere. 2019;**218**:680-686

[110] Belcheva M et al. Antibacterial properties of metal nanoparticles–incorporated activated carbon composites produced from waste biomass precursor. Diamond and Related Materials. 2024;**141**:110545

[111] Nath H et al. Development of lignocellulosic biomass derived Cu and Zn doped highly porous activated carbon and its utilization in the anti-microbial treatment. Environmental Chemistry and Ecotoxicology. 2023;**5**:155-164

[112] Iwar RT, Ugwudike BO. Groundwater fluoride removal by novel activated carbon/aluminium oxide composite derived from raffia palm shells: Optimization of batch operations and field-scale point of use system evaluation. Results in Engineering. 2022;**14**:100407

[113] Kuok K et al. Sustainable clean water production using bamboo activated carbon for rural residents in the Borneo Island. BioResources. 2022;**17**(2):3227

[114] Xing X et al. Enhancing inhibition of disinfection byproducts formation and opportunistic pathogens growth during drinking water distribution by Fe2O3/coconut shell activated carbon. Environmental Pollution. 2021;**268**:115838

[115] Xiao F et al. Production of granular activated carbon by thermal air oxidation of biomass charcoal/biochar for water treatment in rural communities: A mechanistic investigation. Chemical Engineering Journal Advances. 2020;**4**:100035

[116] Chaukura N et al. A new generation low-cost biochar-clay composite 'biscuit'ceramic filter for point-of-use water treatment. Applied Clay Science. 2020;**185**:105409

[117] Suif Z et al. Design of groundwater filter media using activated carbon for emergency purpose. In: Proceedings of the 5th International Conference on Sustainable Civil Engineering Structures and Construction Materials (SCESCM) 2020. Singapore: Springer; 2022:1357-1369

[118] Yusuf K, Murtala M. Development and performance evaluation of a portable household ceramic water filter with activated carbon and magnetic treatment unit. International journal of Environmental Science and Technology. 2020;**17**:4009-4018

[119] Chaffin JD et al. The ability of household pitcher-style water purifiers to remove microcystins depends on filtration rate and activated carbon source. Water Supply. 2019;**19**(1):336-345

[120] Kaipper BI, Madureira LA, Corseuil HX. Use of activated charcoal in a solid-phase extraction technique for analysis of pesticide residues in tomatoes. Journal of the Brazilian Chemical Society. 2001;**12**:514-518

[121] Chen BJ et al. Structural characterization of activated carbon

prepared from passion fruit by-product for removal of cholesterol and toxic compound in liquid food. Journal of Biobased Materials and Bioenergy. 2023;**17**(2):232-244

[122] Khalla D et al. Competitive adsorption of quinary heavy metal ions onto chestnut shell activated carbon. Materials Chemistry and Physics. 2024;**323**:129646

[123] Moi CT, Khan R. Carbon-based absorption materials for heavy metal removal. In: Remediation of Heavy Metals: Sustainable Technologies and Recent Advances. John Wiley & Sons Ltd; 2024. pp. 149-167

[124] Girgis BS, Tmerek YM, Badawy EH. Potential of activated carbon derived from local common reed in the refining of raw cane sugar. Carbon Letters. 2010;**11**(3):192-200

[125] Qureshi K et al. Physical and chemical analysis of activated carbon prepared from sugarcane bagasse and use for sugar decolorisation. International Journal of Chemical and Biomolecular Engineering. 2008;**1**(3):145-149

[126] Mudoga H, Yucel H, Kincal N. Decolorization of sugar syrups using commercial and sugar beet pulp based activated carbons. Bioresource Technology. 2008;**99**(9):3528-3533

[127] Muñoz-González Y et al. Activated carbons from peach stones and pine sawdust by phosphoric acid activation used in clarification and decolorization processes. Journal of Chemical Technology and Biotechnology: International Research in Process, Environmental and Clean Technology. 2009;**84**(1):39-47

[128] Li J et al. Preparation of discarded cigarette butt-derived activated carbon and its decolorization for waste edible oils. Biomass Conversion and Biorefinery. 2023;**13**(11):10299-10309

[129] Rodriguez-Reinoso F. Activated carbon and adsorption. In: Encyclopedia of Materials: Science and Technology. Elsevier Ltd; 2001. pp. 22-34

[130] Jiang Y et al. Ethylene control in fruit quality assurance: A material science perspective. Aggregate. 2024;**5**(5):e565

[131] Chaemsanit S, Matan N, Matan N. Activated carbon for food packaging application. Walailak Journal of Science and Technology (WJST). 2018;**15**(4):255-271

[132] Sirimuangjinda A et al. Production and characterization of activated carbon from waste tire by H3PO4 treatment for ethylene adsorbent used in active packaging. Advanced Materials Research. 2012;**506**:214-217

[133] Ng C et al. Freundlich adsorption isotherms of agricultural by-product-based powdered activated carbons in a geosmin–water system. Bioresource Technology. 2002;**85**(2):131-135

[134] Ahmed MJ. Adsorption of non-steroidal anti-inflammatory drugs from aqueous solution using activated carbons. Journal of Environmental Management. 2017;**190**:274-282

[135] Boudrahem N et al. Single and mixture adsorption of clofibric acid, tetracycline and paracetamol onto activated carbon developed from cotton cloth residue. Process Safety and Environmental Protection. 2017;**111**:544-559

[136] Alvear-Daza JJ, Rengifo-Herrera JA, Pizzio LR. Performance and optimization of diclofenac and ibuprofen adsorption

onto activated carbon synthesized from sunflower seed shell (*Helianthus annuus*) in natural groundwater samples. Adsorption. 2024;**30**:1-18

[137] Bednárek J et al. Revelation of high-adsorption-performance activated carbon for removal of fluoroquinolone antibiotics from water. Biomass Conversion and Biorefinery. 2024;**14**(2):2585-2599

[138] Al-Sareji OJ et al. A sustainable Banana Peel activated carbon for removing pharmaceutical pollutants from different waters: Production, characterization, and application. Materials. 2024;**17**(5):1032

[139] Vinayagam R et al. Magnetic activated carbon synthesized using rubber fig tree leaves for adsorptive removal of tetracycline from aqueous solutions. Environmental Research. 2023;**216**:114775

[140] Liu P et al. The efficient removal of Congo red and ciprofloxacin by peony seeds shell activated carbon with ultra-high specific surface area. Environmental Science and Pollution Research. 2023;**30**(18):53177-53190

[141] Xiong J et al. Biomass-derived mesoporous and super-hydrophilic carbon manufactured by cycling-pressure-switching air activation process towards ultrahigh adsorption efficiency of tetracycline. Sustainable Materials and Technologies. 2022;**32**:e00430

[142] Geng X et al. Carboxyl-functionalized biochar derived from walnut shells with enhanced aqueous adsorption of sulfonamide antibiotics. Journal of Environmental Management. 2021;**280**:111749

[143] Mondal S, Patel S, Majumder SK. Naproxen removal capacity enhancement by transforming the activated carbon into a blended composite material. Water, Air, and Soil Pollution. 2020;**231**(2):37

[144] Chowdhury S et al. Comprehensive analysis on sorptive uptake of enrofloxacin by activated carbon derived from industrial paper sludge. Science of the Total Environment. 2019;**665**:438-452

[145] Sagadevan S et al. Agricultural biomass-based activated carbons for efficient and sustainable supercapacitors. Journal of Energy Storage. 2024;**97**:112878

[146] Chaudhary P et al. Waste biomass-derived activated carbons for various energy storage device applications: A review. Journal of Energy Storage. 2024;**78**:109996

[147] Guo Z et al. Activation of biomass-derived porous carbon for supercapacitors: A review. Chinese Chemical Letters. 2024;**35**(7):109007

[148] Bandara T et al. Activated carbon synthesized from Jack wood biochar for high performing biomass derived composite double layer supercapacitors. Carbon Trends. 2024;**15**:100359

[149] Özçifçi Z et al. Production and characterization of activated carbon foams with various activation agents for electrochemical double layer capacitors (EDLCs) applications. Colloids and Surfaces A: Physicochemical and Engineering Aspects. 2024;**690**:133851

[150] Tekin B, Topcu Y. Novel hemp biomass-derived activated carbon as cathode material for aqueous zinc-ion hybrid supercapacitors: Synthesis, characterization, and electrochemical performance. Journal of Energy Storage. 2024;**77**:109879

[151] Shah SS. Biomass-derived carbon materials for advanced metal-ion hybrid Supercapacitors: A step towards more sustainable energy. Batteries. 2024;**10**(5):168

[152] Ramalingam G et al. Biomass and waste derived silica, activated carbon and ammonia-based materials for energy-related applications–a review. Fuel. 2024;**355**:129490

[153] Ahmed F et al. Fabrication of a biomass-derived activated carbon-based anode for high-performance li-ion batteries. Micromachines. 2023;**14**(1):192

[154] Zheng F et al. Biomass waste inspired nitrogen-doped porous carbon materials as high-performance anode for lithium-ion batteries. Journal of Alloys and Compounds. 2017;**693**:1197-1204

[155] Long W et al. Biomass-derived nanostructured carbons and their composites as anode materials for lithium ion batteries. Chemical Society Reviews. 2017;**46**(23):7176-7190

[156] Sun F et al. A high performance lithium ion capacitor achieved by the integration of a Sn-C anode and a biomass-derived microporous activated carbon cathode. Scientific Reports. 2017;**7**(1):40990

[157] Zhu J et al. Biomass-derived carbons for sodium-ion batteries and sodium-ion capacitors. ChemSusChem. 2020;**13**(6):1275-1295

[158] Hong K-L et al. Biomass derived hard carbon used as a high performance anode material for sodium ion batteries. Journal of Materials Chemistry A. 2014;**2**(32):12733-12738

[159] Molaiyan P et al. Recent progress in biomass-derived carbon materials for Li-ion and Na-ion batteries—A review. Batteries. 2023;**9**(2):116

[160] Dos Reis GS et al. Sustainable biomass activated carbons as electrodes for battery and supercapacitors—A mini-review. Nanomaterials. 2020;**10**(7):1398

[161] Niu J et al. Sustainable synthesis of biomass-derived carbon electrodes with hybrid energy-storage behaviors for use in high-performance Na-ion capacitors. ACS Applied Energy Materials. 2020;**3**(3):2478-2489

[162] Volperts A et al. Biomass based activated carbons for fuel cells. Renewable Energy. 2019;**141**:40-45

[163] Dhelipan M et al. Activated carbon from orange peels as supercapacitor electrode and catalyst support for oxygen reduction reaction in proton exchange membrane fuel cell. Journal of Saudi Chemical Society. 2017;**21**(4):487-494

[164] Balakrishnan P et al. Enhanced performance based on a hybrid cathode backing layer using a biomass derived activated carbon framework for methanol fuel cells. Electrochimica Acta. 2017;**251**:51-59

[165] Sivakami A et al. Porous carbon derived from biomass for fuel cells. In: Biomass-Derived Carbon Materials: Production and Applications. Wiley-VCH GmbH; 2022. pp. 229-252

Chapter 5

Preparation of Biomass Carbon Materials and Their Several Typical Applications

Zhaofeng Wu, Jun Sun, Ping Hu and Weijin Wang

Abstract

In this chapter, "Preparation of Biomass Carbon Materials and Their Several Typical Applications," we delve into the realm of transforming biomass waste into valuable carbon materials. The scope encompasses a comprehensive overview of various biomass sources, highlighting their abundance and renewability as precursors. We explore diverse preparation methods, ranging from traditional pyrolysis and hydrothermal carbonization to more advanced techniques like chemical activation and templating that convert these wastes into high-performance carbon materials. Furthermore, the chapter showcases several typical applications of these biomass-derived carbon materials, emphasizing their potential in energy storage, catalysis, environmental remediation, and even as advanced materials in electronics and sensors. The discussion also touches upon the critical aspect of structure–activity relationships, elucidating how the physicochemical properties of the carbon materials govern their functional performance in different applications. Overall, this chapter aims to provide a holistic understanding of the transformation of biomass waste into valuable carbon materials and their diverse utilization paths, fostering sustainable development and circular economy practices.

Keywords: biomass carbon materials, preparation methods, typical applications, structure-activity relationship, sustainable development

1. Introduction

Carbon materials play a crucial role in the advancement of human civilization and societal development [1–4]. Today, diverse novel carbon materials (NCMs) such as graphene, carbon nanotubes, fullerene, carbon dots, and porous carbon have emerged and are being utilized across various applications due to their versatile structures and properties. However, the primary precursors for these NCMs are nonrenewable and unsustainable resources (coal and petroleum), and their production often involves harsh conditions and high energy consumption, which can lead to environmental pollution. Therefore, it is desirable to use abundant, renewable, and low-cost carbon sources alone with eco-friendly processes for producing NCMs.

Biomass, a renewable energy resource, primarily falls into plant and animal categories, sourced from microbial, crop, and forestry residues, animal carcasses and remains, and municipal solid waste (MSW) (**Figure 1**). And, using biomass, as a precursor for producing NCMs, offers advantages such as low cost, renewability, sustainability, and widespread availability, especially when compared to fossil fuel-based carbon precursors. Biomass consists mainly of carbon (C), hydrogen (H), and oxygen (O), forming carbon networks similar to those in other carbon materials and exhibiting a diverse range of molecular structures. Rationalization of biomass waste as carbon precursors can convert otherwise worthless waste into multifunctional, high-value biomass-derived carbon materials (BCMs). Among the various methods, pyrolysis carbonization and hydrothermal carbonization (HTC) are the two most commonly employed techniques for preparing BCMs [5, 6]. By adjusting parameters such as heating rate and duration, precise control over the structure, composition, and morphology of BCMs can be achieved. Additionally, microwave-assisted carbonization and Joule heating carbonization methods are also utilized in BCMs preparation. Furthermore, activation and templating techniques can significantly influence the porosity of BCMs, allowing for more precise control of the preparation. Similarly, BCMs can be combined with other materials to create or enhance composites, for example, by integrating metal oxide semiconductors (MOSs) to form heterojunctions [7].

Due to their varied structures and different surface chemistries, BCMs have a wide range of potential applications, including energy storage, catalysis, environmental remediation, electronics, and sensors (**Figure 1**). For example, three-dimensional (3D) porous BCMs, with their high porosity, enhance electrolyte diffusion and have many potential applications in energy conversion and memory devices. And, their well-developed pore network also improves the accessibility of active sites, making them effective in catalysis [8]. Additionally, BCMs' high specific surface area (SSA), excellent porosity, and rich surface functional groups make them ideal for use as efficient pollution adsorbents and solar energy steam generators [9]. Moreover, BCMs offer high electrical conductivity, electron mobility, chemical stability, and thermal resilience, positioning them as excellent candidates for flexible electronics and sensor devices [10–15]. In the pursuit of sustainable development and a circular economy, the conversion of biomass waste into valuable carbon materials has emerged as a promising field.

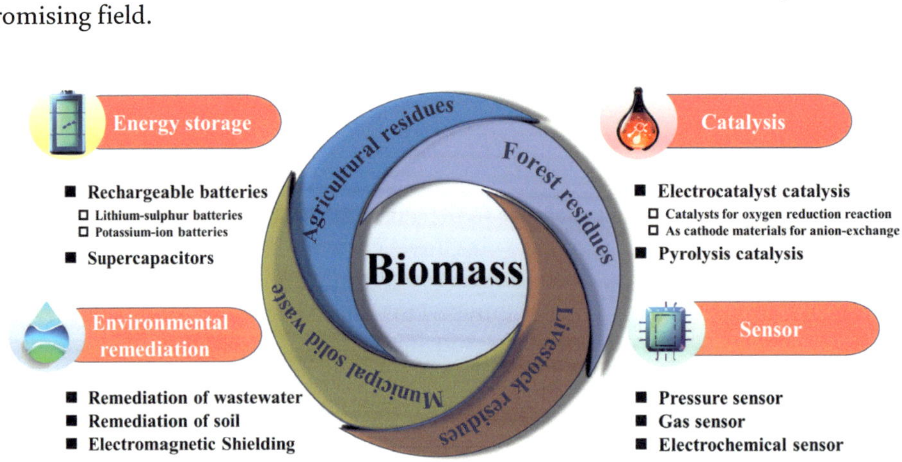

Figure 1.
Classification of BCMs and their applications in various fields.

In the following, this chapter concentrates on the intricate process of transforming biomass, a plentiful renewable resource, into high-performance carbon materials. It begins by describing the various biomass sources and their potential as precursors for carbon material production. Then, it delves into different preparation methods and their impact on the properties of the resulting carbon materials. Finally, it highlights the versatility of these BCMs across a range of applications, from energy storage to environmental remediation.

2. Biomass sources and their carbonization

Biomass is an excellent precursor for producing BCMs due to its abundance and renewability. It encompasses a variety of organic sources, including agricultural, forest, and livestock residues, as well as even MSW, which are classified into plant-derived and animal-derived biomass. Each type has distinct properties that influence the characteristics of the resulting BCMs. For instance, plant biomass is rich in lignocellulose, enhancing the porosity and SSA of the BCMs. In contrast, animal biomass has a more complex composition, so only those animal-derived precursors with easily available proteins and chitins are typically used in BCMs preparation.

2.1 Biomass sources

2.1.1 Agricultural, forest, and livestock residues

Agricultural, forestry, and livestock residues are by-products during the production and management of these industries. They include, but are not limited to, post-harvest straw, rice husks, corncobs, cotton stalks, fruit and vegetable residues, tree trunks, branches, sawdust, wood shavings from timber processing, and animal bones, hides, and manure from livestock production. These residues can be repurposed as BCMs, giving them a new utility. Despite their complex composition, which includes C, H, O, nitrogen (N), phosphorus (P), sulfur (S), and trace metals, these biomass resources can be converted into BCMs through a straightforward process (i.e., carbonization).

2.1.2 MSW

MSW refers to the daily waste discarded by people across cities worldwide. It is usually composed of biodegradable and nonbiodegradable materials from organic and inorganic sources. Annually, approximately 1.9 billion tons of MSW are generated globally, with organic waste and waste paper comprising 46 and 17%, respectively [16]. This suggests that at least 63% of MSW worldwide has the potential to be used as precursors for the creation of multifunctional carbon materials.

2.2 Carbonization of biomass

Carbonization of biomass involves changes in elements' composition with temperature. This process entails the gradual release of H and O as the temperature rises, while C transitions from a complex compound to a single substance. Consequently, temperature is a crucial factor in the production of BCMs. Next, the carbonization process for both plant and animal biomass will be discussed in detail.

2.2.1 Carbonization process of plant-derived biomass

Plants are the most widely studied precursors for BCMs. The main chemical components of agricultural lignocellulosic biomass are cellulose (40–50 wt%), hemicellulose (20–30 wt%), and lignin (10–25 wt%) [17]. **Figure 2a** presents the chemistry of the plant biomass in the cellulose, hemicellulose, and lignin ternary system. The decomposition rate and extent of these components largely depend on the process and parameters, e.g., hydrothermal carbonization, pyrolytic carbonization, and the rate of heating, target temperature, and holding time set during the carbonization process. Cellulose and hemicellulose, with their abundant hydroxyl groups, easily decompose into volatile compounds (CO, CO_2, H_2O, and some hydrocarbons, etc.). At lower temperatures (≤400°C), these processes usually produce many oxygen-containing heterocycles, which are readily converted into aromatic rings by reactions (dehydration, decarboxylation, and decarbonylation) as the temperature increases. The mechanisms of these pyrolysis reactions have been the focus of extensive studies. As shown in **Figure 2b**, hemicellulose decomposes first at 200–350°C, with the maximum mass loss rate (0.95 wt.%/°C) occurring at 268°C. Cellulose decomposes between 315 and 400°C, with peak decomposition rates (2.84 wt.%/°C) at 355°C, and nearly completes decomposition above 400°C. Lignin's decomposition is more gradual, with a wide temperature range (140–900°C) and a lower decomposition rate (<0.14 wt.%/°C) [18]. Additionally, Deng et al. [19] summarized the possible structure revolution in cellulose, hemicellulose, and lignin when heat-treated with $KHCO_3$. And, the numerous hydroxyl groups in cellulose and hemicellulose degrade to form

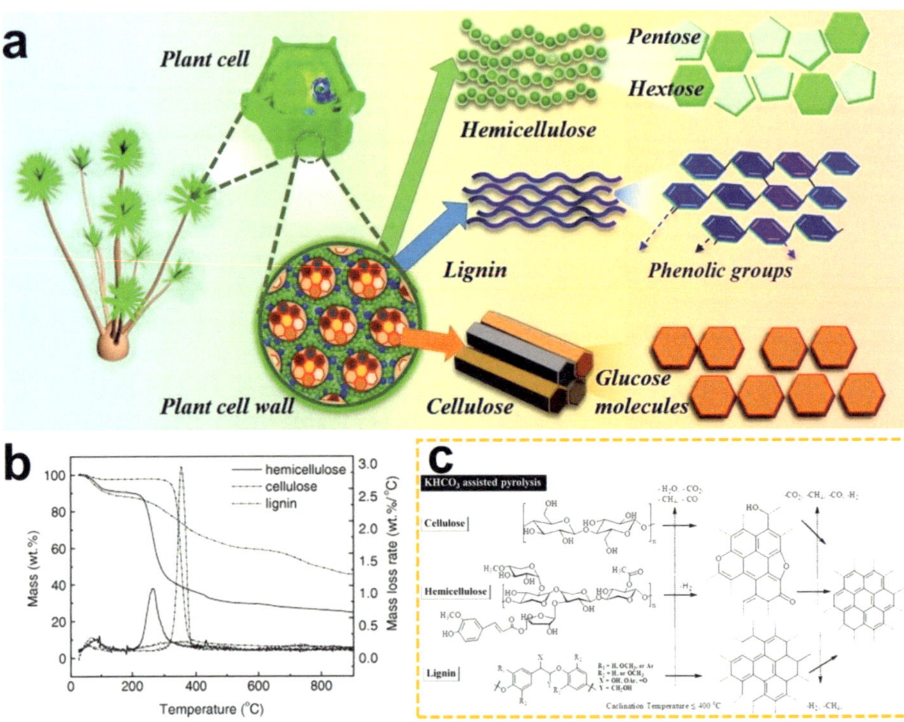

Figure 2.
(a) Ternary system of plant biomass [17]; (b) pyrolysis curves in TGA and DTG [18]; (c) the possible structure revolution under the thermal treatment with the assistance of $KHCO_3$ of cellulose, hemicellulose, and lignin [19].

oxygen-containing heterocycles during carbonization, while lignin, with fewer O and more aromatic rings, converts more readily into carbon materials with fewer micropores (**Figure 2c**).

2.2.2 Carbonization process of animal-derived biomass

Animal biomass contains many complex components, usually rich in chitin, and protein precursors are commonly used for the preparation of BCMs. Chitin, the second most prevalent natural biopolymer, is composed of β-(1 → 4) chitosan-linked *N*-acetyl glucosamine units, which are the *N*-deacetylated product of chitin [20], which can be prepared by pyrolysis in an inert atmosphere (**Figure 3a**). Chitin is thermally stable, with a degradation onset temperature of about 280°C. The main temperature range of degradation is 300–450°C, and the maximum rate of decomposition occurs at 431°C. The 3D FTIR spectra of volatile products are shown in **Figure 3b**, and the pyrolysis of chitin mainly involves the deacetylation and decomposition of deacetylation products [21]. Proteins are biological macromolecules linked by peptide bonds of carious acids [22]. Silk is a typical protein biomass widely used in the preparation of BCMs. He et al. [23] successfully prepared cobalt (Co) and tungsten bimetallic-loaded N-doped porous carbon electrocatalysts derived from silk cellulose by facile carbonization and KCl chemical activation (**Figure 3c**).

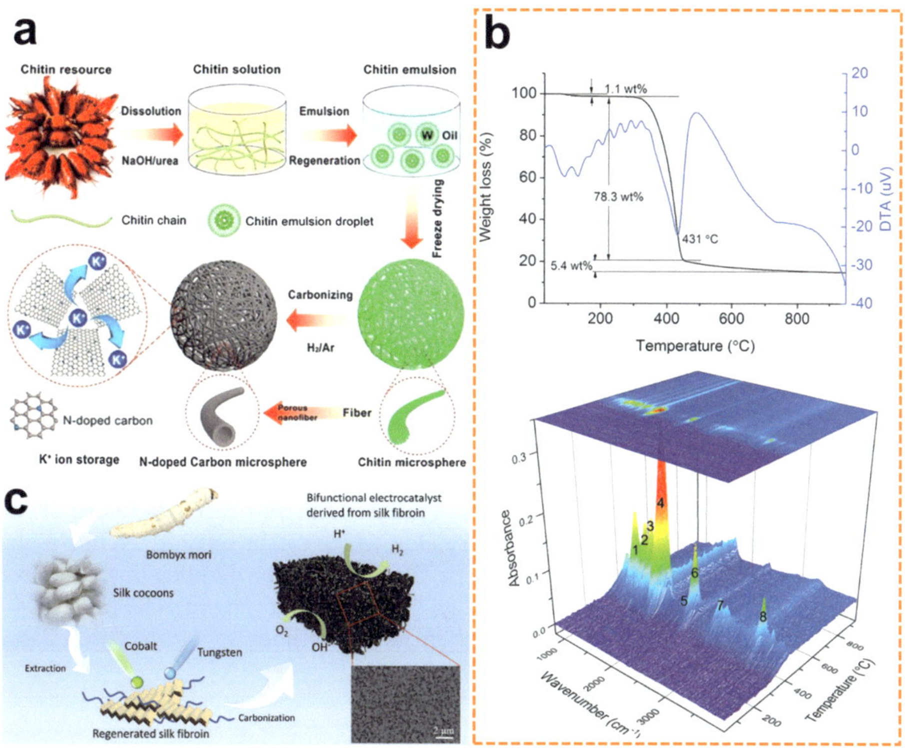

Figure 3.
(a) Direct carbonization of chitin into N-doped materials [24]. (b) TGA and DTA curves of chitin and 3D FTIR spectra of chitin during carbonization [21]. (c) The top of N-doped porous carbon electrocatalysts derived from silk cellulose [23].

3. Preparation methods for BCMs

Biomass carbonization involves heat transfer and numerous chemical reactions, making the choice of carbonization method crucial for altering the physical and chemical properties of the materials. Various carbonization methods impact the morphology of the obtained BCMs. The following discussion will cover several commonly used carbonization methods and explore approaches to enhance the diversity of BCMs.

3.1 Pyrolysis carbonization

Pyrolysis is the most common method for carbonization of biochar. Given the complexity of biomass composition, it varies under different pyrolysis conditions [6]. Pyrolysis methods are typically classified based on temperature rise and residence time into slow pyrolysis, fast pyrolysis, and flash pyrolysis. Slow pyrolysis involves a gradual temperature increase, generally below 10°C/s, which favors the formation of aromatic hydrocarbon structures and preserves structural complexity in biochar [25]. Fast pyrolysis occurs at higher temperatures, between 400 and 650°C, with a heating rate ranging from over 10°C/s to less than 200°C/s. Flash pyrolysis occurs at 800–1000°C, with a heating rate of 100°C/s or even faster. The high heating rate promotes thermal cracking of the biomass, including cracking and hydrolysis, which results in the production of liquids and volatiles, and produces a higher yield of bio-oil [26]. The presence of bio-oil will not be conducive to the contact between the surface of materials and the outside world, reducing the effective SSA. However, in pollutant treatment, particularly in oil–water separation, bio-oil can actually facilitate the separation of BCMs from the liquid, benefiting the subsequent processes.

3.2 HTC

Hydrothermal treatments utilize water as a protective medium during the heating process. Water has an infinite network of hydrogen bonds, which allows the solubility of certain salts to increase under critical conditions, thus facilitating acid or base catalysis [27]. The dissociation of acidic hydrated hydrogen ions and basic hydroxide ions can promote interactions with biomass, break chemical bonds, and facilitate the separation of products and other components, allowing biomass to be cracked at lower temperatures [28]. The hydrothermal process is exothermic producing primarily solid-phase carbon materials, followed by liquid and gaseous phases [29]. Depending on the proportion of the transformation products, hydrothermal treatment can be classed into HTC, hydrothermal liquefaction, and hydrothermal gasification. Excessive liquid phase can reduce SSA and often result in low solid-phase yields in the gas phase. Therefore, HTC is preferred for enhancing solid-phase carbon yields, which can be categorized into high-temperature and low-temperature HTC, with 250–300°C as the dividing line. Moreover, HTC can provide a high-temperature and high-pressure environment, which reduces energy consumption compared to other carbonization methods.

3.3 Other carbonization methods

Additionally, two other prevalent treatments are microwave-assisted carbonization and Joule heating carbonization. Microwave-assisted carbonization, in particular, offers the advantage of achieving greater porosity in a shorter time frame. This method also minimizes the production of hazardous by-products, the use of chemicals, and pollutant

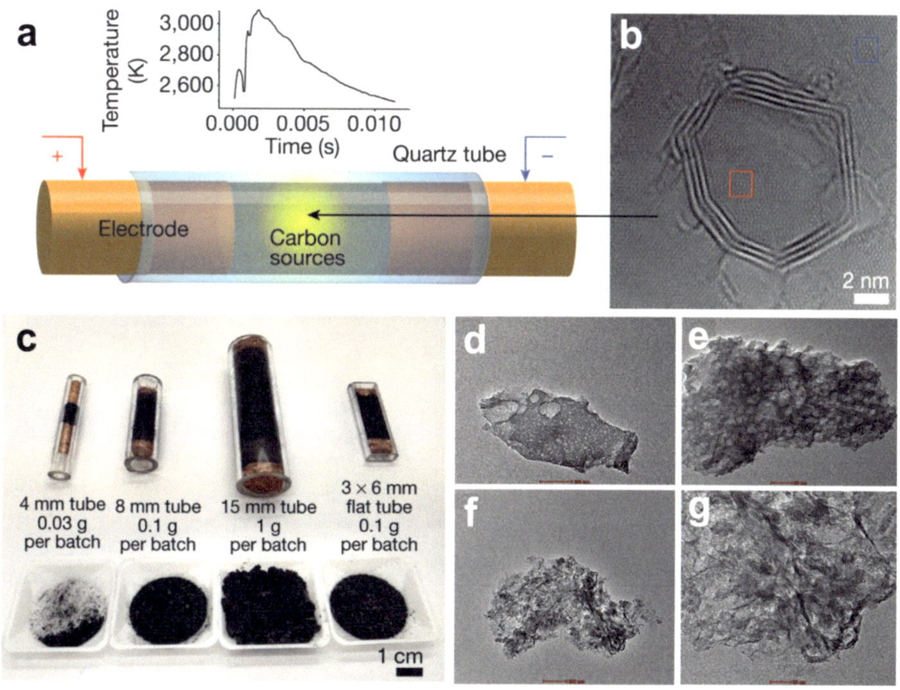

Figure 4.
(a) Schematic diagram of the Joule thermal flash evaporation technology, (b) TEM, and (c) the picutures of its synthesis of graphene samples [32]. (d-g) TEM of the crystalline carbon structure after activation [33].

emissions [30]. However, it has its limitations, including the need for additives to enhance the material's microwave absorption capacity [31]. Furthermore, factors such as the material's dielectric properties, surface shape, and dimensions can lead to uneven heating and make it challenging to control the temperature and rate of heating. Conversely, Joule heating involves heating an electrolyte-filled medium by applying an electric current directly to it. This method heats an electrode or heating rod to high temperatures rapidly, leading to the carbonization of the material. Joule heating has progressively been utilized in the production of BCMs (**Figure 4a–c**). Nevertheless, this technique places significant demands on the equipment and is also prone to uneven heating.

Table 1 summarizes and compares the four carbonization methods discussed. It shows that each method has its prospects (Pros.) and considerations (Cons.), allowing for the selection of the most suitable carbonization approach based on specific needs to achieve the desired BCMs.

3.4 Methods to increase the diversity of BCMs

3.4.1 Activation method

To enhance the diversity and performance of BCMs, activation is crucial for adjusting the structure and performance of biochar. This process is categorized into physical and chemical activation based on the conditions used.

Physical activation occurs entirely in the gas phase and involves two main processes: first, pyrolysis, where the precursor is heated to 400–1000°C in an inert gas environment to eliminate volatiles, which can also produce disorganized carbon that

Method	Parameters	Pros.	Cons.
Pyrolysis carbonization	• Temperature (< 3000°C) • Heating rate • Residence time • Atmosphere	• Wide operating temperature range • Programmable heating • Controllable atmosphere	• Uneven heating • High energy consumption
Hydrothermal carbonization	• Temperature (< 350°C) • Residence time • Doping	• Uniform heating • Easy to control material morphology • Easy doping	• Limited operating temperature range • Low graphitization degree
Microwave-assisted carbonization	• Temperature (< 1000°C) • Residence time • Microwave power	• Noncontact volumetric rapid heating • High safety level	• Difficult temperature and heating rate control • Limited application
Joule heating carbonization	• Temperature (> 3000 K) • Residence time • Voltage and current • Atmosphere	• Rapid heating • High graphitization degree	• Uneven heating • High equipment requirements

Table 1.
Comparison of common carbonization methods.

clogs the pores reduces the SSA; second, gasification, where a physical activator is introduced at 700–1200°C to remove tar and other products from the pores, thereby increasing the porosity, SSA (**Figure 4d-g**), and expose more active sites [34]. The two most commonly used physical activators are carbon dioxide and steam, of which carbon dioxide is popular due to its ability to effectively control the rate of pore transformation and produce graded pores.

Chemical activation leverages kinetically controlled reactions in specialized environments. Activators are typically classified as acidic, basic, or metallic. Acidic and basic activations generally involve mixing with biochar at low temperatures, which enhances the SSA and porosity of BCMs [35]. Additionally, some researchers have developed a composite activation method by combining acid and base treatments [36]. $ZnCl_2$, a common metal activator, acts as a strong dehydrating agent that not only prevents tar formation but also does not react with carbon. It can also facilitate deoxygenation at high temperatures, thereby improving the yield of the final product [37]. Compared to physical activation, chemical activation offers several advantages, including precise control over pore size, a high SSA, a well-distributed microporous structure, elevated carbon content, straightforward operation, rapid reaction rates, lower pyrolysis temperatures, and shorter activation times. However, it also has drawbacks, such as the requirement for extra chemical reagents, challenges with recycling, and the necessity for additional washing steps after use. And, some researchers have proposed combining chemical and physical activation methods to create a hybrid physicochemical activation process [38].

3.4.2 Template-assisted method

The template-assisted synthesis method involves creating inorganic porous materials on the surface of biomass precursors, guiding the formation and growth of BCMs. This approach is typically categorized into hard and soft template methods. Hard templates, such as mesoporous metal oxides or metal-organic frameworks [39], generate mesopores or macropores but require removal and cleaning with an alkaline solution after activation [40]. In contrast, soft templates use surfactants or block copolymers to form mesopores without needing removal after activation [41]. To further refine porosity, researchers are increasingly employing a dual template method that combines both hard and soft templates [42].

4. Typical applications of BCMs

The previous section provided a detailed overview of how BCMs are prepared. This section will explore the latest advancements in their innovative applications across key areas such as energy storage, catalysis, environmental remediation, as well as electronics and sensors.

4.1 Electrochemical energy storage

Given the challenges of renewable energy sources, such as uneven distribution and intermittent availability, energy storage devices have garnered significant interest [43]. As crucial components of these devices (batteries and capacitors), electrode materials play a pivotal role in determining their performance [44]. BCMs, known for their heteroatom doping, porous structures, and high SSA, are viewed as promising candidates with exceptional electrochemical performance [45].

Enhancing the surface wettability of materials through atomic doping and introducing pseudocapacitive charge storage mechanisms represent effective strategies for improving the electrochemical performance of electrodes [46]. He et al. [47] prepared S/N co-doped layered porous BCMs starch through a combination of the template method and calcination (**Figure 5a–b**). These materials exhibited a spongy porous morphology and a layered mesoporous structure with increased interlayer distances, resulting in superior Na storage capacity compared to pure Cui et al. [48] successfully prepared S/N double-doped porous C/SiO_x composites ($SN@C/SiO_x$) using rice husk as the C and Si source, with thiourea for S and N source. When employed as an anode for lithium-ion batteries (LIBs), these composites demonstrated a stable reversible capacity of 1150 mAh/g at a current density of 0.1 A/g, with an initial coulombic efficiency (CE) enhancement of 70.4% (**Figure 5c**). Xu et al. [49] reported on heteroatom-doped mushroom-derived BCMs (MBCMs) for use as an anode material in potassium (K)-ion batteries. The porous MBCMs exhibited a relatively high specific capacity of 290 mAh/g at a current density of 500 mA/g and maintained excellent cycling stability after 2000 cycles (**Figure 5d**). And, combining BCMs with them with transition metal oxides or phosphides, which have strong adsorption capabilities, can further improve their electrochemical performance [50]. Furthermore, BCMs can be activated to achieve a high SSA and a well-defined microporous size distribution. Currently, fish scales [51], waste paper [52], flour [53], yeast cells [54], fallen leaves [55], pig bones [55], willow wadding [55], lettuce leaves [56], waste tea leaves [57], sunflower seed husks [58], ginkgo biloba shells [59], cow dung [60], silk [61], and human hair [62] have been chemically activated for use as supercapacitors.

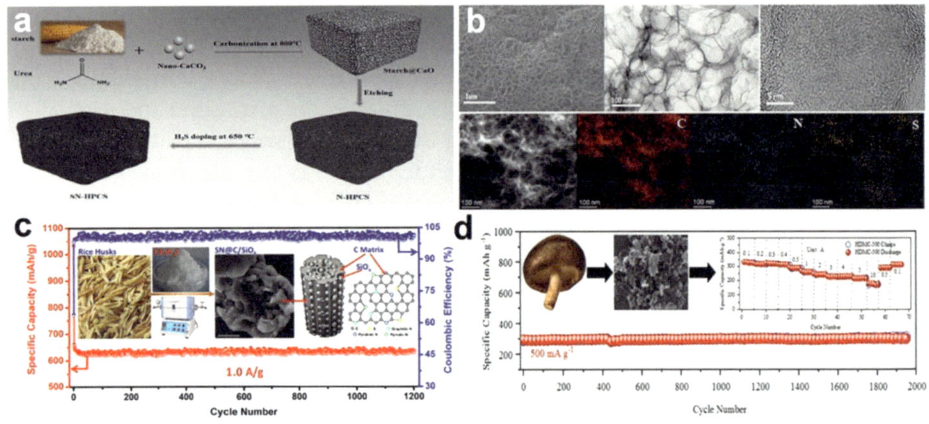

Figure 5.
(a) Schematic illustration of the preparation of SN-HPCS; (b) SEM images, TEM images, HRTEM image, and EDS images of SN-HPCS and the corresponding elemental mapping of C, N, and S elements, respectively [47]; (c) SN@C/SiO$_x$ composites as high-performance LIBs anodes [48]; (d) a MBCMs as high-stability anode for K ion battery [49].

4.2 Catalysis

4.2.1 Electrocatalysis

Employing biomass as a sustainable carbon source, the construction of at 3D multi-stage pore structure, exposure of the active sites, and doping with heteroatoms to modify the electron cloud distribution, thereby enhancing the oxygen reduction reaction (ORR) activity of the catalysts [63]. Kim et al. [64] prepared layered porous S, N-doped bamboo carbon materials SNBCs through the pyrolysis of bamboo and thiourea, which were employed as electrocatalysts for ORR (**Figure 6a**). By precisely controlling the mesopore ratio and optimizing the effective doping amount, the SNBCs achieved half-wave potentials and stability comparable to commercial Pt/C catalysts. Lee et al. [65] successfully synthesized hierarchical 2D porous iron single-atom catalysts using hemoglobin-rich discarded red blood cells (**Figure 6b**). This interesting structure offers excellent mass transfer properties, resulting in superior ORR performance in alkaline media, which was also employed as a cathode material in an anion-exchange membrane fuel cell (AEMFC) and demonstrated a maximum power density of 658 mW/cm^2 (**Figure 6c–d**). Additionally, transition metal compounds loaded on BCMs carriers have good catalytic properties for ORR [66].

4.2.2 Pyrolysis catalysis

Various catalysts have been employed for the catalytic pyrolysis of biomass [67, 68]. Recently, activated carbon has garnered significant attention as an effective catalyst for converting biomass into sustainable, high-quality bio-oil and high-value-added chemicals. Studies have demonstrated that activated carbon derived from corn stover achieves 100% selectivity in glucose-catalyzed pyrolysis of phenolic compounds [69]. Additionally, cellulose conversion was conducted using a catalyst-to-cellulose ratio of 1.13 at a pyrolysis temperature of 450°C, confirming that activated carbon is effective for the pyrolysis of phenolic compounds.

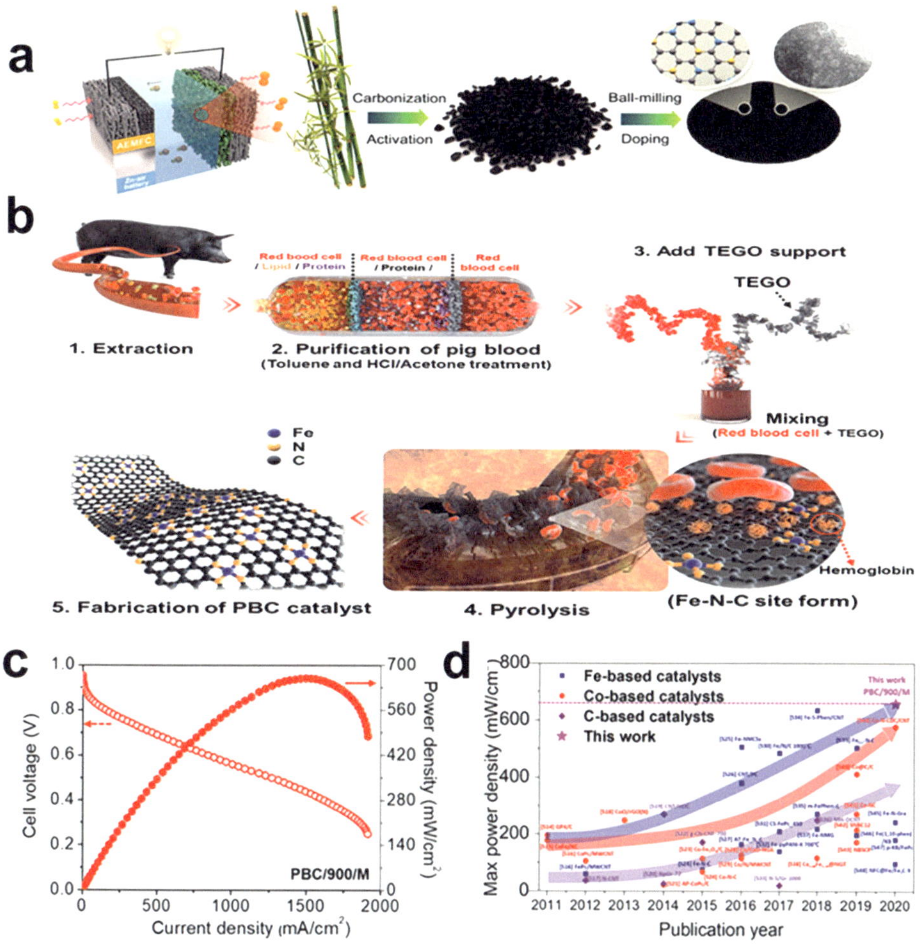

Figure 6.
(a) Preparation and application of SNBCs [64]; (b) schematic illustration of the synthesis for PBC catalysts; (c) single-cell polarization result obtained using PBC/900/M as the cathode catalyst with a 4 mg/cm² loading at 1 bar pressure. The anode was 0.6 mg/cm² PtRu/C, and a 60 wt% ionomer was used; (d) comparison of the best AEMFC performance using developed catalysts with time [65].

4.3 Environmental remediation

4.3.1 Remediation of wastewater

BCMs are mainly utilized as an affordable, eco-friendly filter for removing harmful pollutants from wastewater and stormwater. Among various options, cellulose nanofibers stand out as excellent adsorbents due to their unique properties, abundance, and renewability. Lu et al. [70] detailed the adsorption mechanisms of lead (Pb^{2+}) on sludge biochar surfaces, identifying four potential processes (**Figure 7a**): (a) electrostatic outer surface complexation; (b) internal complexation and co-precipitation with metals oxides and organics; (c) surface complexation with the -OH and -COOH functional groups present; and (d) metal precipitation as Pb-PbO-silicates. Consequently, biochar and biochar-based filtration technologies are promising options for wastewater treatment and discharge.

4.3.2 Remediation of soil

BCMs exert varied effects on the mobility of metals in soil. Kong et al. [71] reported a removal efficiency of 86.4% for Hg (II) and 99.5% for phenanthrene using soybean straw biochar (**Figure 7b**). This high adsorption performance is attributed to its significant microporosity and SSA. In the remediation of inorganic pollutants, BCMs primarily function through mechanisms such as precipitation, electrostatic attraction, and ion exchange. Notably, the pyrolysis temperature significantly influences the physiochemical properties of biochar, which in turn affects its ability to adsorb both organic and inorganic contaminants. Cao et al. [72] also investigated the effect of biochar on Pb fixation in soil, finding that P in the biochar could immobilize Pb by forming insoluble hydroxy pyrophosphate.

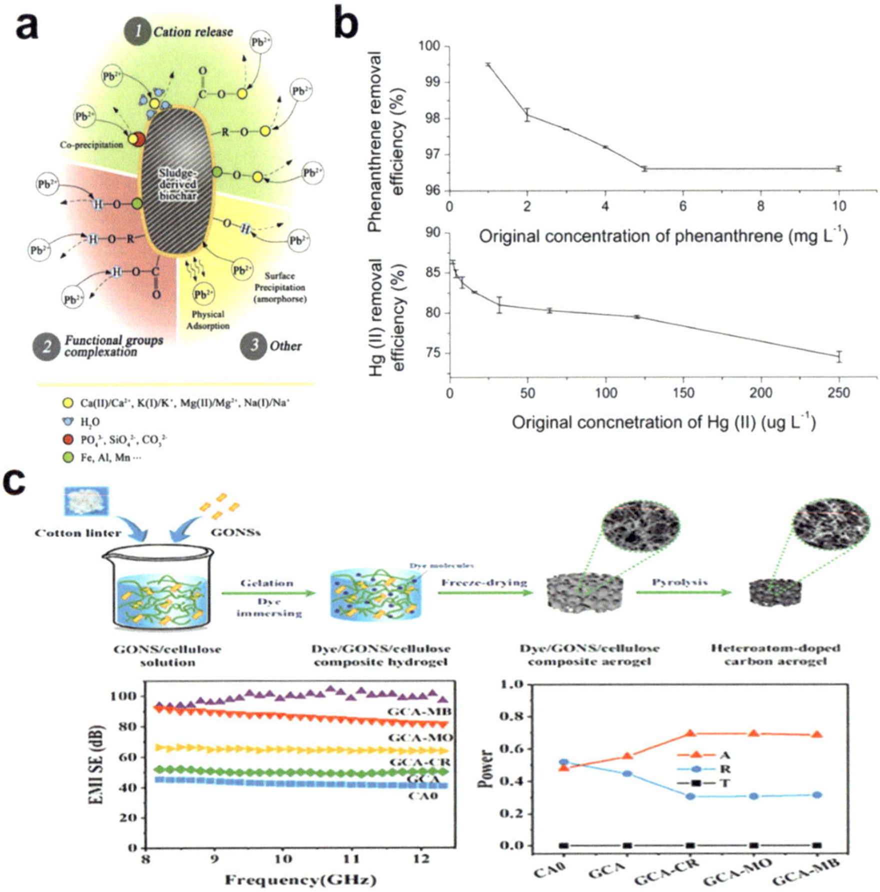

Figure 7.
(a) The adsorption mechanism of sludge biochar [70]; (b) the removal efficiencies of phenanthrene and Hg (II) from solution by the soybean straw-derived biochar [71]; (c) a novel heteroatom-doped cellulose-derived carbon aerogel with hierarchically porous architecture and hybrid conductive network exhibits an outstanding performance [73].

4.3.3 Electromagnetic shielding

In recent years, electromagnetic interference (EMI) pollution has become a growing concern, significantly increasing the demand for effective shielding materials. Wang et al. [73] prepared heteroatom-doped carbon aerogels for organic dye adsorption by synthesizing graphene oxide nanosheets/cellulose composite hydrogels, followed by freeze-drying and a specific two-stage carbonization process (**Figure 7c**). Despite their low density of 68.9 mg/cm^3, these aerogels demonstrated an EMI shielding effectiveness of up to 97.3 dB and an absorption coefficient of up to 0.69 in the X-band. BCMs offer a promising new approach for developing high-performance wave-absorbing materials [6, 74].

4.4 Electronics and sensors

BCMs feature a wide range of structural and morphological designs, offer straightforward functionalization, and have adjustable surface and electronic properties [75]. Currently, BCMs are employed in various electronic devices, encompassing e-textiles as well as applications in compound, metal ion, gas detection, and humidity–temperature sensing [76].

4.4.1 Pressure sensor

Electronic skin is utilized in various modern scientific fields. Guo's group prepared a low-cost, efficient, and highly sensitive flexible tactile sensor using calathea zebrine (CZ) [77] and lotus leaf (LF) [78] as templates. Guo et al. [77] synthesized a micro-structured ionic gel (MIG) using CZ leaves as a template using soft lithography. This MIG features a lower detection limit (LoD) of 0.1 Pa for high-performance capacitive electronic skin. Additionally, an electric double layer (EDL) forms at both the top and bottom interfaces, with the EDL area at the rough interface increasing under cone compression. The polydimethylsiloxane (m-PDMS) structure, replicated using LF, includes high aspect ratio, low-density micro-towers, and the bottom electrodes covered with ultrathin silver nanowires [78]. The flexible sensor boasts a LoD under 0.8 Pa, a response time of less than 36 ms, and remains fatigue-resistant after 100,000 touches. Additionally, BCM composites combined with other materials offer promising applications in flexible sensing research [79, 80].

4.4.2 Gas sensor

The rich pore structure of BCMs provides effective diffusion channels for the adsorption and desorption of target gasses. Their extensive SSA also offers more accessible active sites for gas detection, enabling the sensors to identify low gas concentrations at room temperature (RT), overcoming the problems that traditional gas sensors have failed to overcome in real life, and has attracted great attention.

Among the methods available, the biomimetic approach has emerged as a key technique for fabricating nanoscale structured materials. Biomass templates can be classified according to natural biological species: plant templates (e.g., wood [81, 82], bamboo [83], straw [84], etc.); animal templates (e.g., eggshell membranes [85, 86], crab shells [87, 88], insect wings [89], etc.); microbial templates (e.g., fungi [90], yeast, viruses, etc.); and biomolecular templates (e.g., proteins, DNA, etc.).

Liu et al. [91] developed porous nickel-Co oxide (Ni-Co-O) nanocomposites by adjusting the Co-Ni ratio of initial Ni-Co nitrate, utilizing hemp stems as a template.

The resulting sensor exhibited a high response (3 ~ 100 ppm), low LoD (50 ppb), and stability (>7 weeks) for NH_3 at RT. Shao et al. [92] prepared porous ZnO-graded microtubes by zinc nitrate impregnation and air calcination using spent rose stems as bio templates, examining how different calcination temperatures affected microstructure and gas-sensitive properties. The ZnO-500 sensor exhibited a short response time (2 s), a high response (S = 143.0), and a low LoD (10 ppb) to 50 ppm n-butanol vapor at an operating temperature of 252°C. These gas-sensitive indicators were significantly better than those of the already reported ZnO-based n-butanol sensors. Additionally, many natural materials with multi-level pore structures, which feature two or more pore channels, can be used to obtain porous biochar through simple carbonization for gas-sensing applications [93, 94].

The striated folds present on the surface of roses, forming micro/nanostructures, and the intricate multistage pore architectures within discarded bamboo chopsticks, characterized by interconnected pores spanning various levels and specialized for achieving a high specific surface area, exhibit structural similarities to the nasal turbinates of dogs. Drawing inspiration from the intrinsic micro/nanostructure of biomass, Sun et al. [93] pioneered the conversion of discarded rose tea into a biomass carbon with a striped folded structure akin to that of the dog's maxillary nasal turbinate. This material demonstrated exceptional performance, including high responsiveness, a low detection limit, swift recovery (2.0 s), and long-term stability toward NH_3 at ambient temperatures. This biomimetic sensor has been effectively employed for real-time lamb freshness monitoring (**Figure 8a**). In parallel, Qin et al. [95] achieved the carbonization of discarded disposable bamboo chopsticks into a high specific surface area carbon material, replicating the structure of a dog's nasal turbinate under various temperature conditions (**Figure 8b**). Building upon this foundation, Qin et al. [7] further developed tin dioxide (SnO_2) heterojunctions on the surface of discarded bamboo chopsticks through a straightforward one-step pyrolytic carbonization process. The bamboo chopsticks were transformed into a tubular porous structure, mimicking the dog's nasal turbinate, while SnO_2 served as an olfactory receptor for the recognition and detection of NH_3 at RT (**Figure 8c**).

In recent years, biochar composites have garnered increasing attention in the field of gas sensing. Chen et al. [96] and his team used hemp stem biochar (BC) as a carrier and synthesized CoBC composite material using MOF template through vacuum-assisted and calcining methods. The CoBC-700 composite showed an excellent response to 100 ppm NO_2 gas (R_a/R_g = 34.1), showing impressive reproducibility and stability at RT. Sun et al. [97] introduced hollow porous carbon microtubules (CMTs) and ZnO/CMTs composites with heterojunctions by simply carbonizing sycamore villus fibers. Both CMTs and ZnO/CMTs exhibit higher sensitivity and faster response (<16 s) and recovery time (<2 s), along with good long-term stability. Zhang et al. [98] synthesized imine covalent organic scaffolds (iCOF) at RT using micron tube structured BCMs from lavender straw as a template, and this groundbreaking study is the first to combine COF materials with biological carbon materials. This sensor achieved a remarkable maximum response of 2155.75% to aniline, one total circle time under 26 s, and a LoD of 10.26 ppb. These advancements are likely to inspire further research into high-quality, biomass-derived gas-sensitive materials.

4.4.3 Other electronics

Lu et al. [99] developed a BCMs doped with both N and P from lotus root using a one-step pyrolysis method and created an electrochemical sensor for detecting baicalein (BA) and luteolin (LU) from this material. The characterization results

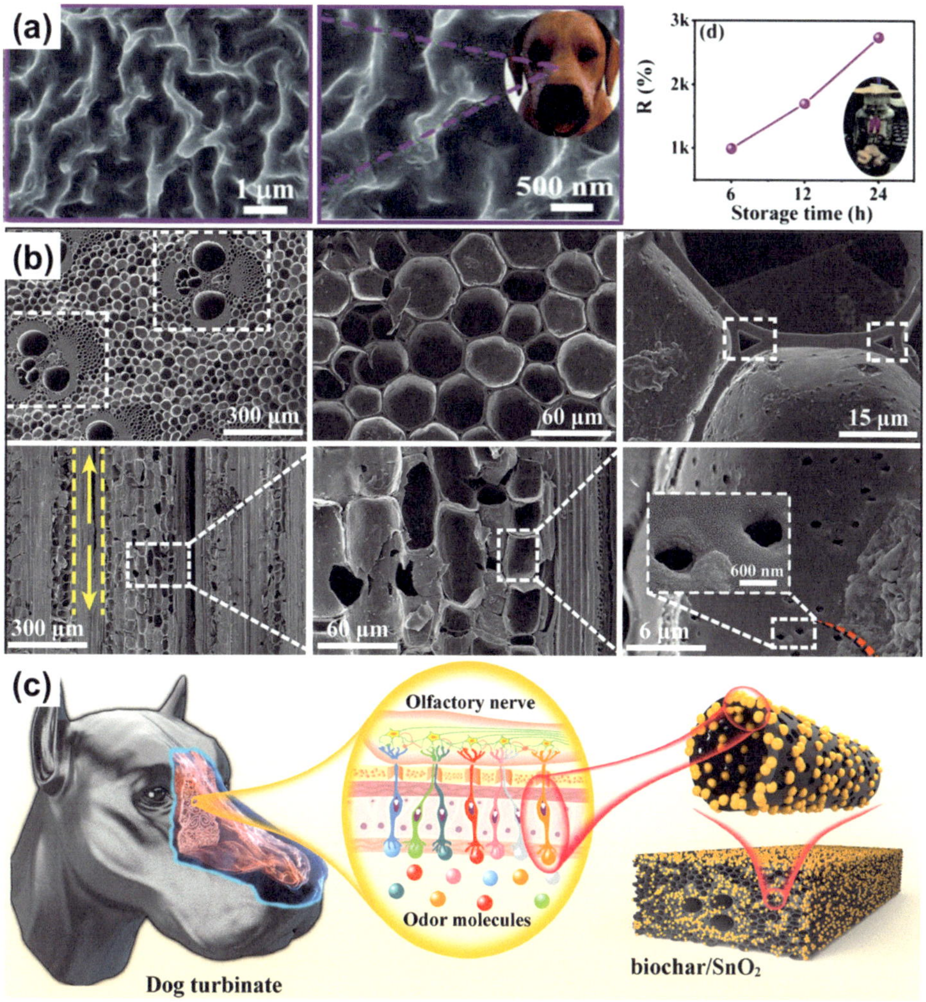

Figure 8.
(a) SEM images with different resolutions of CRT and lamb stored at RT for 6, 12, and 24 h the response line graphs [93]; (b) longitudinal sections and transverse sections SEM of DBC–600 [95]; (c) biomimetic schematic of biochar/SnO_2 [7].

indicate that a precise amount of elemental doping significantly enhances the SSA of the carbon material. However, excessive doping can drastically reduce this SSA. In addition, co-doping with N and P effectively increases the number of active sites and improves the electron transfer rate of the materials. Under optimal detection conditions, the sensor achieved minimum detection limits of 7.8×10^{-9} mol/L for BA and 7.6×10^{-9} mol/L for LU (S/N = 3). The sensor also demonstrated strong performance in detecting these compounds in both Shuang Huang Lian oral solutions and urine samples from healthy individuals.

5. Conclusions

In summary, this chapter explores the structures of various BCMs and provides an in-depth discussion of notable synthetic methods and their features. It also reviews recent progress in biomass materials for diverse applications. Although progress has been made in the research of BCMs, there are still many challenges to be addressed. Firstly, preparing BCMs necessitates a thorough understanding of the precursors' microstructure and an assessment of their pore size and dimensions. Secondly, there are currently no guidelines for choosing the appropriate biochar to achieve effective heteroatom doping, which is necessary to enhance interlayer spacing, introduce defects, and create active sites. Additionally, selecting the ideal shapes and gradations of pores from their precursors remains challenging. Then, when chemical activators are employed to achieve high SSA and hierarchical porous structures, obtaining large pore sizes can be challenging. Additionally, controlling pore geometry, size, and connectivity remains difficult. Finally, controlling the proportion of impurity atoms introduced during doping modifications is challenging. At all, it is hoped that the energy savings discussed in this chapter will encourage further research and development of BCMs, supporting sustainable development and circular economy practices.

Acknowledgements

This research was funded by Xinjiang Natural Science Fund for Distinguished Young Scholars (2022D01E37), Tianshan Innovation Team Program of Xinjiang Uygur Autonomous Region (2023D14001) and Tianshan Talent Training Project-Xinjiang Science and Technology Innovation Team Program (2023TSYCTD0012).

Conflict of interest

The authors declare no conflict of interest.

Author details

Zhaofeng Wu[1,2]*, Jun Sun[1,3], Ping Hu[1,3] and Weijin Wang[1,2]

1 Xinjiang Key Laboratory of Solid-State Physics and Devices, Urumqi, Xinjiang, China

2 School of Materials Science and Engineering, Xinjiang University, Urumqi, China

3 School of Physics Science and Technology, Xinjiang University, Urumqi, Xinjiang, China

*Address all correspondence to: wuzf@xju.edu.cn

References

[1] Liu B, Wang X, Chen Y, Xie H, Zhao X, Nassr AB, et al. Honeycomb carbon obtained from coal liquefaction residual asphaltene for high-performance supercapacitors in ionic and organic liquid-based electrolytes. Journal of Energy Storage. 2023;**68**:107826. DOI: 10.1016/j.est.2023.107826

[2] Liu YY, Shi HD, Wu ZS. Recent status, key strategies and challenging perspectives of fast-charging graphite anodes for lithium-ion batteries. Energy & Environmental Science. 2023;**16**(11):4834-4871. DOI: 10.1039/D3EE02213G

[3] Zhong M, Zhang M, Li X. Carbon nanomaterials and their composites for supercapacitors. Carbon Energy. 2022;**4**(5):950-985. DOI: 10.1002/cey2.219

[4] Dong X, Liu X, Chen H, Xu X, Hu Y. Hard template-assisted N, P-doped multifunctional mesoporous carbon for supercapacitors and hydrogen evolution reaction. Journal of Materials Science. 2021;**56**(3):1-14. DOI: 10.1002/cey2.219

[5] Kan T, Strezov V, Evans TJ. Lignocellulosic biomass pyrolysis: A review of product properties and effects of pyrolysis parameters. Renewable and Sustainable Energy Reviews. 2016;**57**:1126-1140. DOI: 10.1016/j.rser.2015.12.185

[6] Jiang Y, Wang W, Liang M, Lai X, Xie K, Liao L, et al. 3D hierarchical porous composites based on natural loofahs/cobalt nanoparticles/graphene oxide with efficient electromagnetic interference shielding performance. Journal of Alloys and Compounds. 2021;**889**:161716. DOI: 10.1016/j.jallcom.2021.161716

[7] Qin Z, Wu Z, Sun Q, Sun J, Zhang M, Chen F, et al. Dog nose-inspired high-performance ammonia sensor based on biochar/SnO_2 composite. Carbon. 2023;**213**:118297. DOI: 10.1016/j.carbon.2023.118297

[8] Luo H, Kaneti YV, Ai Y, Wu Y, Wei F, Fu J, et al. Nanoarchitectured porous conducting polymers: From controlled synthesis to advanced applications. Advanced Materials. 2021;**33**(29):2007318. DOI: 10.1002/adma.202007318

[9] Mohamad Nor N, Lau LC, Lee KT, Mohamed AR. Synthesis of activated carbon from lignocellulosic biomass and its applications in air pollution control-A review. Journal of Environmental Chemical Engineering. 2013;**1**(4):658-666. DOI: 10.1016/j.jece.2013.09.017

[10] Wang C, Xia K, Wang H, Liang X, Yin Z, Zhang Y. Advanced carbon for flexible and wearable electronics. Advanced Materials. 2019;**31**(9):1801072. DOI: 10.1002/adma.201801072

[11] You J, Li M, Ding B, Wu X, Li C. Crab chitin-based 2D soft nanomaterials for fully biobased electric devices. Advanced Materials. 2017;**29**(19):1606895. DOI: 10.1002/adma.201606895

[12] Li Q, Zhang LN, Tao XM, Ding X. Temperature sensors: Review of flexible temperature sensing networks for wearable physiological monitoring. Advanced Healthcare Materials. 2017;**6**(12):1601371. DOI: 10.1002/adhm.201770059

[13] Afify AS, Ahmad S, Khushnood RA, Jagdale P, Tulliani JM. Elaboration and characterization of novel humidity sensor based on micro-carbonized

bamboo particles. Sensors and Actuators B: Chemical. 2017;**239**:1251-1256. DOI: 10.1016/j.snb.2016.09.130

[14] Zhong X, Yuan W, Kang Y, Xie J, Hu F, Li CM. Biomass-derived hierarchical nanoporous carbon with rich functional groups for direct-electron-transfer-based glucose sensing. ChemElectroChem. 2016;**3**(1):144-151. DOI: 10.1002/celc.201500351

[15] Wang H, Li S, Wang Y, Wang H, Shen X, Zhang M, et al. Bioinspired fluffy fabric with in situ grown carbon nanotubes for ultrasensitive wearable airflow sensor. Advanced Materials. 2020;**32**(11):1908214. DOI: 10.1002/adma.201908214

[16] Nanda S, Berruti F. A technical review of bioenergy and resource recovery from municipal solid waste. Journal of Hazardous Materials. 2021;**403**:123970. DOI: 10.1016/j.jhazmat.2020.123970

[17] Anwar Z, Gulfraz M, Irshad M. Agro-industrial lignocellulosic biomass a key to unlock the future bio-energy: A brief review. Journal of Radiation Research and Applied Sciences. 2014;**7**(2):163-173. DOI: 10.1016/j.jrras.2014.02.003

[18] Yang H, Yan R, Chen H, Lee DH, Zheng C. Characteristics of hemicellulose, cellulose and lignin pyrolysis. Fuel. 2007;**86**(12):1781-1788. DOI: 10.1016/j.fuel.2006.12.013

[19] Deng J, Xiong T, Wang H, Zheng A, Wang Y. Effects of cellulose, hemicellulose, and lignin on the structure and morphology of porous carbons. ACS Sustainable Chemistry & Engineering. 2016;**4**(7):3750-3756. DOI: 10.1021/acssuschemeng.6b00388

[20] Khan I, Ali N, Jing Z, Khan A, Ali F, Hhan F, et al. Biopolymer-carbonaceous composites, progress, and adsorptive mitigation of water pollutants. International Journal of Biological Macromolecules. 2024;**274**:133379. DOI: 10.1016/j.ijbiomac.2024.133379

[21] Gao Y, Chen X, Zhang J, Yan N. Chitin-derived mesoporous, nitrogen-containing carbon for heavy-metal removal and styrene epoxidation. ChemPlusChem. 2015;**80**(10):1556-1564. DOI: 10.1002/cplu.201500293

[22] Cho SY, Yun YS, Lee S, Jang D, Park KY, Kim JK, et al. Carbonization of a stable β-sheet-rich silk protein into a pseudographitic pyroprotein. Nature Communications. 2015;**6**(1):7145. DOI: 10.1038/ncomms8145

[23] He H, Zhang Y, Zhang W, Li Y, Wang Y, Wang P, et al. Dual metal-loaded porous carbon materials derived from silk fibroin as bifunctional electrocatalysts for hydrogen evolution reaction and oxygen evolution reaction. ACS Applied Materials & Interfaces. 2021;**13**(26):30678-30692. DOI: 10.1021/acsami.1c07058

[24] Chen C, Wang Z, Zhang B, Miao L, Cai J, Peng L, et al. Nitrogen-rich hard carbon as a highly durable anode for high-power potassium-ion batteries. Energy Storage Materials. 2017;**8**:161-168. DOI: 10.1016/j.ensm.2017.05.010

[25] Bandara YW, Gamage P, Gunarathne DS. Hot water washing of rice husk for ash removal: The effect of washing temperature, washing time and particle size. Renewable Energy. 2020;**153**:646-652. DOI: 10.1016/j.renene.2020.02.038

[26] Yaashikaa PR, Senthil Kumar P, Varjani SJ, Saravanan A. Advances in production and application of biochar from lignocellulosic feedstocks for remediation of environmental

pollutants. Bioresource Technology. 2019;**292**:122030. DOI: 10.1016/j.biortech.2019.122030

[27] Byrd AJ, Kumar S, Kong L, Ramsurn H, Gupta RB. Hydrogen production from catalytic gasification of switchgrass biocrude in supercritical water. International Journal of Hydrogen Energy. 2011;**36**(5):3426-3433. DOI: 10.1016/j.ijhydene.2010.12.026

[28] Aysu T, Abd Rahman NA, Sanna A. Catalytic pyrolysis of tetraselmis and isochrysis microalgae by nickel ceria based catalysts for hydrocarbon production. Energy. 2016;**103**:205-214. DOI: 10.1016/j.energy.2016.02.055

[29] Zhang Z, Zhu Z, Shen B, Liu L. Insights into biochar and hydrochar production and applications: A review. Energy. 2019;**171**:581-598. DOI: 10.1016/j.energy.2019.01.035

[30] Ao W, Fu J, Mao X, Kang Q, Ran C, Liu Y, et al. Microwave assisted preparation of activated carbon from biomass: A review. Renewable and Sustainable Energy Reviews. 2018;**92**:958-979. DOI: 10.1016/j.rser.2018.04.051

[31] Mushtaq F, Mat R, Ani FN. A review on microwave assisted pyrolysis of coal and biomass for fuel production. Renewable and Sustainable Energy Reviews. 2014;**39**:555-574. DOI: 10.1016/j.rser.2014.07.073

[32] Luong DX, Bets KV, Algozeeb WA, Stanford MG, Kittrell C, Chen W, et al. Gram-scale bottom-up flash graphene synthesis. Nature. 2020;**577**(7792):647-651. DOI: 10.1038/s41586-020-1938-0

[33] Divya P, Rajalakshmi R. Facile synthesis of micro/mesoporous functional carbon from Turbinaria conoides seaweed for high performance of supercapacitors. Materials Today: Proceedings. 2020;**27**:44-53. DOI: 10.1016/j.matpr.2019.08.200

[34] Rashidi NA, Yusup S. Potential of palm kernel shell as activated carbon precursors through single stage activation technique for carbon dioxide adsorption. Journal of Cleaner Production. 2017;**168**:474-486. DOI: 10.1016/j.jclepro.2017.09.045

[35] Hai A, Bharath G, Babu KR, Taher H, Naushad M, Banat F. Date seeds biomass-derived activated carbon for efficient removal of NaCl from saline solution. Process Safety and Environmental Protection. 2019;**129**:103-111. DOI: 10.1016/j.psep.2019.06.024

[36] Feng Y, Lu H, Liu Y, Xue L, Dionysiou DD, Yang L, et al. Nano-cerium oxide functionalized biochar for phosphate retention: Preparation, optimization and rice paddy application. Chemosphere. 2017;**185**:816-825. DOI: 10.1016/j.chemosphere.2017.07.107

[37] Huang Y, Liu Z, Zhao G. Reaction process for $ZnCl_2$ activation of phenol liquefied wood fibers. RSC Advances. 2016;**6**(82):78909-78917. DOI: 10.1039/C6RA15705J

[38] Hadi S, Taheri E, Amin MM, Fatehizadeh A, Aminabhavi TM. Adsorption of 4-chlorophenol by magnetized activated carbon from pomegranate husk using dual stage chemical activation. Chemosphere. 2021;**270**:128623. DOI: 10.1016/j.chemosphere.2020.128623

[39] Luo D, Han P, Shi L, Huang J, Yu J, Lin Y, et al. Biomass-derived nitrogen/oxygen co-doped hierarchical porous carbon with a large specific surface area for ultrafast and long-life sodium-ion batteries. Applied Surface Science.

2018;**462**:713-719. DOI: 10.1016/j.apsusc.2018.08.106

[40] Li J, Jiang Q, Wei L, Zhong L, Wang X. Simple and scalable synthesis of hierarchical porous carbon derived from cornstalk without pith for high capacitance and energy density. Journal of Materials Chemistry A. 2020;**8**(3):1469-1479. DOI: 10.1039/C9TA07864A

[41] Hosseini H, Behbahani M, Mahyari M, Kazerooni H, Bagheri A, Shaabani A. Ordered carbohydrate-derived porous carbons immobilized gold nanoparticles as a new electrode material for electrocatalytical oxidation and determination of nicotinamide adenine dinucleotide. Biosensors and Bioelectronics. 2014;**59**:412-417. DOI: 10.1016/j.bios.2014.02.046

[42] Wen X, Zhang D, Yan T, Zhang J, Shi L. Three-dimensional graphene-based hierarchically porous carbon composites prepared by a dual-template strategy for capacitive deionization. Journal of Materials Chemistry A. 2013;**1**(39):12334-12344. DOI: 10.1039/C3TA12683H

[43] Yuan X, Zhu B, Feng J, Wang C, Cai X, Qin R. Recent advance of biomass-derived carbon as anode for sustainable potassium ion battery. Chemical Engineering Journal. 2021;**405**:126897. DOI: 10.1016/j.cej.2020.126897

[44] Yang Z, Ren J, Zhang Z, Chen X, Guan G, Qiu L, et al. Recent advancement of nanostructured carbon for energy applications. Chemical Reviews. 2015;**115**(11):5159-5223. DOI: 10.1021/cr5006217

[45] He Y, Zhuang X, Lei C, Lei L, Hou Y, Mai Y, et al. Porous carbon nanosheets: Synthetic strategies and electrochemical energy related applications. Nano Today. 2019;**24**:103-119. DOI: 10.1016/j.nantod.2018.12.004

[46] Yang Z, Gao Y, Zhao Z, Wang Y, Wu Y, Wang X. Phytic acid assisted formation of P-doped hard carbon anode with enhanced capacity and rate capability for lithium ion capacitors. Journal of Power Sources. 2020;**474**:228500. DOI: 10.1016/j.jpowsour.2020.228500

[47] He L, Sun W, Sun K, Mao Y, Deng T, Fang L, et al. Nitrogen and sulfur co-doped hierarchically mesoporous carbon derived from biomass as high-performance anode materials for superior sodium storage. Journal of Power Sources. 2022;**526**:231019. DOI: 10.1016/j.jpowsour.2022.231019

[48] Cui J, Qiu Y, Zhang H, Yao Z, Zhao W, Liu Y, et al. Sulfur- and nitrogen-doped rice husk-derived C/SiO_x composites as high-performance lithium-ion battery anodes. Solid State Ionics. 2021;**361**:115548. DOI: 10.1016/j.ssi.2021.115548

[49] Xu L, Gong Z, Zhang C, Li N, Tang Z, Du J. A mushroom derived biomass carbon as high-stability anode for potassium ion battery. Journal of Alloys and Compounds. 2023;**934**:167820. DOI: 10.1016/j.jallcom.2022.167820

[50] Song Y, Wang J, Li X, Zhao C, Huo J, He C. Boosting polysulfides immobilization and conversion through CoS_2 catalytic sites loaded carbon fiber for robust lithium sulfur batteries. Journal of Colloid and Interface Science. 2022;**608**:963-972. DOI: 10.1016/j.jcis.2021.10.015

[51] Chen W, Zhang H, Huang Y, Wang W. A fish scale based hierarchical lamellar porous carbon material obtained using

a natural template for high performance electrochemical capacitors. Journal of Materials Chemistry. 2010;**20**(23):4773-4775. DOI: 10.1039/C0JM00382D

[52] Kalpana D, Cho SH, Lee SB, Lee YS, Misra R, Renganathan NG. Recycled waste paper-A new source of raw material for electric double-layer capacitors. Journal of Power Sources. 2009;**190**(2):587-591. DOI: 10.1016/j.jpowsour.2009.01.058

[53] Wu X, Jiang L, Long C, Fan Z. From flour to honeycomb-like carbon foam: Carbon makes room for high energy density supercapacitors. Nano Energy. 2015;**13**:527-536. DOI: 10.1016/j.nanoen.2015.03.013

[54] Sun H, He W, Zong C, Lu L. Template-free synthesis of renewable macroporous carbon via yeast cells for high-performance supercapacitor electrode materials. ACS Applied Materials & Interfaces. 2013;**5**(6):2261-2268. DOI: 10.1021/am400206r

[55] Li YT, Pi YT, Lu LM, Xu SH, Ren TZ. Hierarchical porous active carbon from fallen leaves by synergy of K_2CO_3 and their supercapacitor performance. Journal of Power Sources. 2015;**299**:519-528. DOI: 10.1016/j.jpowsour.2015.09.039

[56] Wang R, Wang P, Yan X, Lang J, Peng C, Xue Q. Promising porous carbon derived from celtuce leaves with outstanding supercapacitance and CO_2 capture performance. ACS Applied Materials & Interfaces. 2012;**4**(11):5800-5806. DOI: 10.1021/am302077c

[57] Peng C, Yan XB, Wang RT, Lang JW, Ou YJ, Xue QJ. Promising activated carbons derived from waste tea-leaves and their application in high performance supercapacitors electrodes. Electrochimica Acta. 2013;**87**:401-408. DOI: 10.1016/j.electacta.2012.09.082

[58] Li X, Xing W, Zhuo S, Zhou J, Li F, Qiao SZ, et al. Preparation of capacitor's electrode from sunflower seed shell. Bioresource Technology. 2011;**102**(2):1118-1123. DOI: 10.1016/j.biortech.2010.08.110

[59] Jiang L, Yan J, Hao L, Xue R, Sun G, Yi B. High rate performance activated carbons prepared from ginkgo shells for electrochemical supercapacitors. Carbon. 2013;**56**:146-154. DOI: 10.1016/j.carbon.2012.12.085

[60] Bhattacharjya D, Yu JS. Activated carbon made from cow dung as electrode material for electrochemical double layer capacitor. Journal of Power Sources. 2014;**262**:224-231. DOI: 10.1016/j.jpowsour.2014.03.143

[61] Hou J, Cao C, Idrees F, Ma X. Hierarchical porous nitrogen-doped carbon nanosheets derived from silk for ultrahigh-capacity battery anodes and supercapacitors. ACS Nano. 2015;**9**(3):2556-2564. DOI: 10.1021/nn506394r

[62] Qian W, Sun F, Xu Y, Qiu L, Liu C, Wang S, et al. Human hair-derived carbon flakes for electrochemical supercapacitors. Energy & Environmental Science. 2014;**7**(1):379-386. DOI: 10.1039/C3EE43111H

[63] Liu S, Wang A, Liu Y, Zhou W, Wen H, Zhang H, et al. Catalytically active carbon for oxygen reduction reaction in energy conversion: Recent advances and future perspectives. Advanced Science. 2024;**11**(22):2308040. DOI: 10.1002/advs.202308040

[64] Kim MJ, Park JE, Kim S, Lim MS, Jin A, Kim OH, et al. Biomass-derived air cathode materials: Pore-controlled S,N-Co-doped carbon for fuel cells and metal-air batteries. ACS Catalysis. 2019;**9**(4):3389-3398. DOI: 10.1021/acscatal.8b03730

[65] Lee J, Kim HS, Jang J-H, Lee E-H, Jeong H-W, Lee K-S, et al. Atomic-scale engineered Fe single-atom electrocatalyst based on waste pig blood for high-performance AEMFCs. ACS Sustainable Chemistry & Engineering. 2021;**9**(23):7863-7872. DOI: 10.1021/acssuschemeng.1c01590

[66] Jing B, You S, Ma Y, Xing Z, Chen H, Dai Y, et al. Fe_3Se_4/FeSe heterojunctions in cornstalk-derived N-doped carbon framework enhance charge transfer and cathodic oxygen reduction reaction to boost bio-electricity generation. Applied Catalysis B: Environmental. 2019;**244**:465-474. DOI: 10.1016/j.apcatb.2018.11.074

[67] Anex RP, Aden A, Kazi FK, Fortman J, Swanson RM, Wright MM, et al. Techno-economic comparison of biomass-to-transportation fuels via pyrolysis, gasification, and biochemical pathways. Fuel. 2010;**89**:S29-S35. DOI: 10.1016/j.fuel.2010.07.015

[68] Saraçoğlu E, Uzun BB, Apaydın-Varol E. Upgrading of fast pyrolysis bio-oil over Fe modified ZSM-5 catalyst to enhance the formation of phenolic compounds. International Journal of Hydrogen Energy. 2017;**42**(33):21476-21486. DOI: 10.1016/j.ijhydene.2017.07.001

[69] Zhang Y, Lei H, Yang Z, Duan D, Villota E, Ruan R. From glucose-based carbohydrates to phenol-rich bio-oils integrated with syngas production via catalytic pyrolysis over an activated carbon catalyst. Green Chemistry. 2018;**20**(14):3346-3358. DOI: 10.1039/C8GC00593A

[70] Lu H, Zhang W, Yang Y, Huang X, Wang S, Qiu R. Relative distribution of Pb^{2+} sorption mechanisms by sludge-derived biochar. Water Research. 2012;**46**(3):854-862. DOI: 10.1016/j.watres.2011.11.058

[71] Kong H, He J, Gao Y, Wu H, Zhu X. Cosorption of phenanthrene and mercury(II) from aqueous solution by soybean stalk-based biochar. Journal of Agricultural and Food Chemistry. 2011;**59**(22):12116-12123. DOI: 10.1021/jf202924a

[72] Cao X, Ma L, Liang Y, Gao B, Harris W. Simultaneous immobilization of lead and atrazine in contaminated soils using dairy-manure biochar. Environmental Science & Technology. 2011;**45**(11):4884-4889. DOI: 10.1021/es103752u

[73] Wang ML, Zhang S, Zhou ZH, Zhu JL, Gao JF, Dai K, et al. Facile heteroatom doping of biomass-derived carbon aerogels with hierarchically porous architecture and hybrid conductive network: Towards high electromagnetic interference shielding effectiveness and high absorption coefficient. Composites Part B: Engineering. 2021;**224**:109175. DOI: 10.1016/j.compositesb.2021.109175

[74] Li J, Ding Y, Yu N, Gao Q, Fan X, Wei X, et al. Lightweight and stiff carbon foams derived from rigid thermosetting polyimide foam with superior electromagnetic interference shielding performance. Carbon. 2020;**158**:45-54. DOI: 10.1016/j.carbon.2019.11.075

[75] Srisonphan S. Nanogaps mediated field effect-controlled field emission triode. ACS Applied Electronic Materials. 2022;**4**(10):4857-4862. DOI: 10.1021/acsaelm.2c00790

[76] Xu Y, Lei W, Zhang Y, Fan H, Hao Q, Gao S. Bamboo fungus-derived porous nitrogen-doped carbon for the fast, sensitive determination of bisphenol A. Journal of the Electrochemical Society. 2016;**164**(5):B3043. DOI: 10.1149/2.0021705jes

[77] Qiu Z, Wan Y, Zhou W, Yang J, Yang J, Huang J, et al. Ionic skin

with biomimetic dielectric layer templated from calathea zebrine leaf. Advanced Functional Materials. 2018;**28**(37):1802343. DOI: 10.1002/adfm.201802343

[78] Wan Y, Qiu Z, Hong Y, Wang Y, Zhang J, Liu Q, et al. A highly sensitive flexible capacitive tactile sensor with sparse and high-aspect-ratio microstructures. Advanced Electronic Materials. 2018;**4**(4):1700586. DOI: 10.1002/aelm.201700586

[79] Huang J, Li D, Zhao M, Lv P, Lucia L, Wei Q. Highly stretchable and bio-based sensors for sensitive strain detection of angular displacements. Cellulose. 2019;**26**(5):3401-3413. DOI: 10.1007/s10570-019-02313-3

[80] Fu D, Wang R, Yang R. A novel highly fold-tolerant edgewo-rthia chrysantha lindi-derived substrate for flexible pressure sensors. Cellulose. 2022;**29**(4):2539-2551. DOI: 10.1007/s10570-022-04460-6

[81] Gao X, Zhang S, Du W, Gong X, Nguyen TT, Guo M, et al. Wood-inspired high-performing hierarchical porous $Ce_{0.7}Zr_{0.3}O_2$ catalyst for low-temperature selective catalytic reduction of NO_x by NH_3. Ceramics International. 2021;**47**(20):29149-29161. DOI: 10.1016/j.ceramint.2021.07.078

[82] Yang T, Liu Y, Xia G, Zhu X, Zhao Y. Degradation of formaldehyde and methylene blue using wood-templated biomimetic TiO_2. Journal of Cleaner Production. 2021;**329**:129726. DOI: 10.1016/j.jclepro.2021.129726

[83] Pandoli OG, Neto RJG, Oliveira NR, Fingolo AC, Corrêa CC, Ghavami K, et al. Ultra-highly conductive hollow channels guided by a bamboo bio-template for electric and electrochemical devices. Journal of Materials Chemistry A. 2020;**8**(7):4030-4039. DOI: 10.1039/c9ta13069a

[84] Li B, Liu H, Zeng Q, Dong S, Feng W. Hierarchical porous NiO doped ZnO nanocomposite for formaldehyde gas sensor with high sensitivity, fast response/recovery and good selectivity. Surfaces and Interfaces. 2023;**36**:102502. DOI: 10.1016/j.surfin.2022.102502

[85] Li M, Yan W, Zhu H, Xia S, Wu H, Tang Z. Synthesis and gas sensing properties of biomorphic SnO_2 derived from loofah sponge and eggshell membrane. Journal of Materials Science: Materials in Electronics. 2015;**26**(12):9561-9570. DOI: 10.1007/s10854-015-3619-6

[86] Fan S, Zhao M, Ding L, Liang J, Chen J, Li Y, et al. Synthesis of 3D hierarchical porous Co_3O_4 film by eggshell membrane for non-enzymatic glucose detection. Journal of Electroanalytical Chemistry. 2016;**775**:52-57. DOI: 10.1016/j.jelechem.2016.05.035

[87] Li X, Liu J, Sun M, Sha T, Bo X, Zhou M. Amperometric sensing of ascorbic acid by using a glassy carbon electrode modified with mesoporous carbon nanorods. Mikrochimica Acta. 2018;**185**(10):474. DOI: 10.1007/s00604-018-3010-4

[88] Wu J, Xia M, Zhang X, Chen Y, Sun F, Wang X, et al. Hierarchical porous carbon derived from wood tar using crab as the template: Performance on supercapacitor. Journal of Power Sources. 2020;**455**:227982. DOI: 10.1016/j.jpowsour.2020.227982

[89] Imrith MK, Rosunee S, Unmar R. Bio-inspired knitted fabric development using 3D modelling and image processing. The Journal of the Textile Institute. 2019;**111**(8):1123-1139. DOI: 10.1080/00405000.2019.1686880

[90] Joy F, Devasia J, Nizam A, Lakshmaiah VV, Krishna SBN. Fungi-templated silver nanoparticle composite: Synthesis, characterization, and its applications. Applied Sciences. 2023;**13**(4):2158. DOI: 10.3390/app13042158

[91] Liu H, Lv H, Kan K, Liu Y, Zhang W, Wang Y, et al. Biocarbon-templated synthesis of porous Ni-Co-O nanocomposites for room-temperature NH_3 sensors. New Journal of Chemistry. 2018;**42**(21):17606-17614. DOI: 10.1039/c8nj03832e

[92] Shao TC, Li C, Wang N, Zhang XF, Deng ZP, Xu YM, et al. Biomass-derived porous ZnO hierarchical microtubules for conductometric detection of n-butanol vapor. Sensors and Actuators B: Chemical. 2021;**344**:130293. DOI: 10.1016/j.snb.2021.130293

[93] Sun Q, Wu Z, Qin Z, Chen X, Zhang C, Cao B, et al. A dog nose-inspired high-performance NH_3 gas sensor of biomass carbon materials with a pleated structure derived from rose tea. Journal of Materials Chemistry A. 2022;**10**(27):14326-14335. DOI: 10.1039/D2TA02670H

[94] Peng C, Li S, Wang Y, Ge L, Zhang S, Cai Q, et al. Preparation of Er-Nd-TiO_2 nanocomposite for the highly selective enrichment of phosphotyrosine peptides. Microchemical Journal. 2023;**190**:108597. DOI: 10.1016/j.microc.2023.108597

[95] Qin Z, Wu Z, Sun Q, Sun J, Zhang M, Shaymurat T, et al. Biomimetic gas sensor derived from disposable bamboo chopsticks for highly sensitive and selective detection of NH_3. Chemical Engineering Journal. 2023;**462**:142203. DOI: 10.1016/j.cej.2023.142203

[96] Chen J, Lv H, Bai X, Liu Z, He L, Wang J, et al. Synthesis of hierarchically porous Co_3O_4/biomass carbon composites derived from MOFs and their highly NO_2 gas sensing performance. Microporous and Mesoporous Materials. 2021;**321**:111108. DOI: 10.1016/j.micromeso.2021.111108

[97] Sun Q, Wu Z, Zhang M, Qin Z, Cao S, Zhong F, et al. Improved gas-sensitive properties by a heterojunction of hollow porous carbon microtubes derived from sycamore fibers. ACS Sustainable Chemistry & Engineering. 2021;**9**(43):14345-14352. DOI: 10.1021/acssuschemeng.1c02956

[98] Zhang W, Wang W, Liu S, Tian N, Sun Q, Wu Z. Highly sensitive detection toward aniline vapor promoted by heterojunction between covalent organic frameworks and biocarbon. Applied Surface Science. 2024;**649**:159114. DOI: 10.1016/j.apsusc.2023.159114

[99] Lu Y, Mu X, Liu Y, Gao Y, Shi Z, Zheng Y, et al. The N&P co-doped lotus root derived carbon materials for highly sensitive electrochemical analysis of baicalein and luteolin. Electroanalysis. 2023;**35**(1):215-223. DOI: 10.1002/elan.202200129

Chapter 6

Technoeconomic Review of Briquette Production in a Screw Press [SP] and Hydraulic Piston Press [HPP]

Raphael Segun Bello, Abel Olajide Olorunnisola, Temidayo Emmanuel Omoniyi and Musiliu Ademuiwa Onilude

Abstract

The increase in energy usage over the past century has raised concern over the energy insecurity and environmental unsustainability of current fossil fuel utilization; therefore, there is a need for energy diversification. An attractive alternative diversification is the densification of the abundant biomass. However, poor performance of raw biomass in energy generation necessitates attention to the process through which they were produced. This chapter reviews the technoeconomic impacts of extrusion and compaction, by evaluating two types of extruders: screw press [SP] and hydraulic piston press [HPP] machines. Technical evaluations based on both biomass material and briquetting technologies were reviewed. The biomass technical review includes material availability, pretreatments, and characteristics. The technicalities around the briquetting technologies reviewed include operational variables, machine performance variables, efficiency of the process, and power consumption patterns, while the economic analysis reviewed consists of the cost of biomass, briquetting processes, and market values of final products. From the review, the technical analysis showed that SP has improved the physical quality and combustion characteristics compared with piston press. Two critical challenges in the development of screw press are that of alignment of the extrusion head with the die and high cost of electricity tariff, which discouraged the availability of small-scale screw presses. The economic analysis of briquetting process revealed a considerable cost and end-user savings in HPP briquetting over screw press at the expense of energy value of final products. The economic viability of each process based on different factors of production depended on the material availability, technology deployed, and energy consumption in the production of unit products. The energy cost was a major barrier to achieving sustainability for biomass briquetting with screw press technology. Operating the screw press is not economically viable because of the high electricity tariffs, poor electricity supplies, and high level of machine precisions required. However, the use of alternative energy facilities such as diesel engine as source of power and die heating has reduced the cost of briquette production to 50%.

Keywords: technoeconomic, extrusion, compaction, briquette, extrusion performance, product characteristics

1. Introduction

The current population growth and increase in global energy demands have placed a lot of pressure on fossil fuel, the most exploited energy resource, which contributes about 80% of the primary energy requirement [1–4]. This global energy need is consequential on the outrageous increases in fuel prices in developing countries of Africa, Nigeria, inclusive, and Asia [2]. The situation is worst in Nigeria, where reliance on non-renewable energy resources and capacity to maintain a stable energy stock is very low, with increasing volatile energy markets [5, 6]. The global increase in fossil fuel prices, coupled with the worsening effects of global warming, is edging toward creating a global energy crisis and mitigating climate change. Within the context of environmental protection policy provisions and strategic execution of development plans, biomass is an indispensable part of alternative energy resource debates. As the society becomes more mindful of their influence on the environment, quest for eco-friendly and environmentally sustainable alternatives to biomass is slowly becoming one of the fastest growing affordable renewable technology material sources, with consumption increasing by 3% per year [7, 8].

In creating further awareness toward the exploitation of biomass for renewable energy production, production of biomass-derived energy carriers such as bio-electricity and biofuels could translate to an efficient means of meeting the United Nations Sustainable Development Goals (SDGs) on Affordable and Clean Energy [9]. The current energy insecurity and unsustainable environmental consequences are surmountable through energy resource diversification. The global strive to find alternative routes for energy production led to the implementation of a shared regulatory framework to promote renewable energies, thereby diversifying the sources of supply through the replacement of biomass resources with more viable energy-value products [6]. This has led to the emergence of briquettes as viable alternative energy sources as factored through deviation from the present sustenance energy-usage level to more sustainable and diversified energy options [4].

Technological innovations to increase productivity and reduce costs remained the major challenge to expanding the share of renewable energy carriers, and efforts should be directed toward the development of more user-friendly, cost-efficient technologies at various scales to attract more investment in the field [9]. Therefore, it is necessary to design systems such as briquetting that are capable of obtaining valuable products and drastically reducing the costs of agro-waste and wood residue recovery. Briquetting, otherwise known as densification, is a process by which agricultural and other organic residues are compacted into high-density products with improved energy values compared to the raw materials [10, 11]. Briquetting converts loose materials into uniform-shaped solid fuels (briquettes) with improved handling characteristics, increased volumetric calorific value, reduced transportation costs, and are made available for a variety of applications. A review of various briquetting technologies shows that successful briquetting is dependent on conditions of the material biomass characteristics, the technology adopted, and the variability of briquetting process conditions.

Biomass material characteristics: Parameters identified in the literature that extremely influence the performance characteristics of briquettes are material variables and non-material process variables [12]. Briquette's final quality is affected by three parameters: material characteristics (variables); process variables; and operational variables [13, 14]. Material-specific variables include biomass particle size, particle moisture, and binder. The non-material variables are components of process variables. A significant inference is that structural and material characteristics of biomass are vital to the biomass conversion process and evaluation. Kipyego [15] reported the significance of particle moisture, heating value, ash content, and alkali content as important quality parameters for densification technologies. However, biomass thermochemical characterization (i.e., combustion, gasification, and pyrolysis) is based on the determination of the overall product's physical, mechanical, and chemical characteristics [15]. Variability in material structural characteristics has brought about differences in applicable technological processes and machines used in briquette production.

Technology adopted: Biomass densification employs different agglomeration technologies for the conversion of loose biomass residues into solid fuel, signifying the technology directions. Research studies presented in **Table 1** show some technical characteristics of briquetting technology. Conversely, a screw press has better overall production performance and product quality than piston presses.

Compaction and extrusion processing occurred in two stages: elastic and plastic deformations [24]. During compaction, elastic deformation occurred as the materials regained their original position at pressure relief while the material permanently deformed during plastic deformation. These materials are produced in low-, medium-, or high-pressure processes. In the screw press extrusion process, high-pressure application through a conical screw shaft within an enclosed barrel forced materials through an opening, forcing conveyed material into uniform regular or irregular cross-sections identical to cylindrical bars or hollow tubes with that of the opening [25]. Due to high compressibility and biomass material factors, briquettes produced usually have high-density factors requiring no binders. The preliminary results on machine production showed failed extrusion; no briquette was extruded. The recovered briquettes around the die neck are poorly formed. However, their physical and mechanical attributes showed good comparative characteristics with varying degrees of fractures, carbonation, and good crushing resistance when subjected to impact force. It was observed that there was discontinuity in material flow from the barrel into the die at the transition zone, hence causing deformation and leading to failed extrusion.

Briquetting process conditions: Briquetting process conditions are optimum briquetting conditions that could affect the physical and combustion characteristics of briquettes. Several successful attempts have been made to produce briquettes by varying process conditions (non-material variables), which include the material moisture, the operating temperature, operating pressure, dwell/hold time, feed rate, die geometry, and other machine-specific variables depending on the technology adopted [26–30]. Other process conditions exist that could affect briquette quality, which include material blowout, pre-densification charring, localized heating, high temperatures within the transport/compression sections of the extrusion chamber, and moisture variations as the system equilibrates [31].

1.1 Biomass availability and utilization

Replacement of fossil energy carriers with biomass-derived energy carriers could bring about positive impacts from economic, environmental, and health perspectives. Moreover, production of biomass-derived energy could be achieved under any geographical conditions owing to the large global availability of biomass [32–36]. These wastes are not utilized efficiently, in most cases, are burnt in the open fields resulting not only in energy wastage but also in environmentally polluting manner. The strong position of the abundance of renewable resources in promoting the use of renewable energies in developing countries is a strong factor

Briquette press	Reference	Output capacity	Shape and dimension	Material used	Briquette characteristics
Screw extruder	Sen et al. [16]	120 kg/h	Hexagonal 100 mm length	Cassava rhizome waste	Density: 0.69–0.91 g/cm^3, compressive strength: 8.51–14.94 kg/cm^2, shattering index: 153.7–416.7, HHV: 21,670–24,367 KJ/kg.
	Hamid et al. [17]	200 kg/h	Hexagonal 50 mm length, 20 mm int. dia	Rubber seed kernel (RSK), Palm oil shell (POS)	POS maximum compressive load 101.11 N, HHV: 16.05 MJ/kg RSK briquette compressive load 141 N, HHV: 16.03 MJ/kg.
Mechanical piston press	Srivastava et al. [18]	500 kg/h	Cylindrical. 50 mm dia.	Vegetable market waste (VMW)	Density: 509–747 kg/m^3, from initial densities of 44.2–60 kg/m^3. Briquettes HHV ranges 10.26–16.60 MJ/kg.
	Gill et al. [19]	1200 kg/h	Cylindrical. 70 mm	Rice straw	Density: (1030.38–1159.22 kg/m^3), durability: (71.9–92.3%), HHV: 15.61 MJ/kg, ash content: 16.34%.
Hydraulic Piston Press [HPP]	Moreno et al. [20]	Not available	Cylindrical. 50 mm dia.	Furniture, wood waste, foam.	Briquettes generated more energy and durability, similar to furniture wood waste.
	Manyuchi et al. [21]	Not available	Rectangular. 30 mm length, 25 mm width, 15 mm height	Sawdust, coal fines	HHV: 26 MJ/kg, fixed carbon: 76%, and compressive strength: 0.25 kN/cm^2
Roller press	Thoreson et al. [22]	Not available	Almond-shaped. 31.3 mm length, 23.3 mm width, 17.9 mm depth.	Corn stover, Switchgrass	Bulk densities: 351–527 kg/m^3, durability: 39–90%, crushing strengths: 28–277 N.
	Borowski et al. [23]	Not available	Pillow shaped. 60 mm width, 50 mm height, 30 mm depth.	Charcoal powder	Briquettes' physical properties are satisfactory and suitable for barbecues.

Source: Kpalo et al. [12].

Table 1.
Literature summary of briquette presses study.

influencing growth and economic development [37]. The global biomass resources largely revolve around wastes generated from forestry residue and agricultural and industrial/domestic wastes.

The global supply from different sources was estimated at approximately 220 billion tons per year, with an estimated two-thirds consumed in developing nations in household cooking [38]. This huge diversity of biomass has shown that these materials are abundantly cheap and available, and their conversion from direct burning to generate heat is part of the solution to the world's energy supply and alleviation of environmental problems [4]. Considering the huge potentials in the exploitation of biomass as an energy source in scaled-up plants for heat and power cogeneration, the use of biomass has led to the emergence of mega biodiesel and ethanol production plants [9, 39]. However, there are issues around the global supply of biomass toward the expansion of the biodiesel industry and suitability [4, 40]. The exigency of disposing of these large volumes of waste and other relatively homogeneous wood-based industrial wastes has led to the growth of the briquetting industry.

Despite the advantages of biomass utilization as a viable replacement for fossil fuels for energy production, there are constraints on its direct use as primary biomass for energy production [41]. These constraints, such as a luminous yellow flame, poor energy per unit volume, hydrophilic properties, and high levels of smoke generation above acceptable levels for household use [4, 41, 42], make its utilization and replacement as fossil fuels for energy production difficult [42]. One way by which these wastes can be utilized significantly is by biomass upgrade before utilization [3, 43].

1.2 Technological solutions to biomass upgrade (biomass pretreatment)

Viable technological solutions proffered in literature for biomass property upgrade to meet such demands include biomass pretreatment and fermentation. Biomass pretreatment for briquetting has existed for a long time and is an important process for loosening or delignification of the compacted biomass structures to expose the cellulose fiber [44, 45]. This process removes the volatile matter contents of the lignocellulosic materials, thereby improving their densification, physical, mechanical, and combustion properties [46]. Due to its impact on technical, economic, and environmental systems, biomass pretreatment has drawn significant study attention worldwide [47].

Pretreatment of biomass aimed at delignifying compacted biomass structures to expose the cellulose fiber [44, 45, 48]. This material process could equally increase the production rate up to 340–360 kg/hr [49]. Kpalo et al. [2] revealed that the concept of preheating biomass material in a screw press resulted in a total average energy saving of about 10.2% (23.5% at the heater and 10.8% at the motor). Various pretreatment methods have been identified as relevant to material improvement and sustainable alternative fuel, including thermal (preheating), chemical (fermentation), biological, and mechanical processes, as well as their combinations [50, 51].

The thermal preheating process produces high-quality and high-energy biomass at reduced power consumption [4]. Due to its impact on technical, economic, and environmental systems, biomass pretreatment has drawn significant study attention worldwide [4, 47, 52, 53], showing various thermal conversion processes, routes, and their potential high-value by-products (**Figure 1**) [53]. Thermal preheating offers increased production rates up to 360 kg/hr. and reduced compression and extrusion pressures up to 3.0×10^4 kN/m^2 compared to 1.8×10^5 kN/m^2 without preheating [46, 49, 54, 55]. Kpalo et al. [2] revealed that the concept of preheating biomass material in a screw press showed a total average energy saving of about 10.2% (23.5% at the heater and 10.8% at the motor).

Thermal pretreatment such as torrefaction and pyrolysis occurs under different temperatures and environmental conditions. Torrefaction is the carbonization of the hemicellulose, depolymerization of lignin and cellulose, and devolatilization of

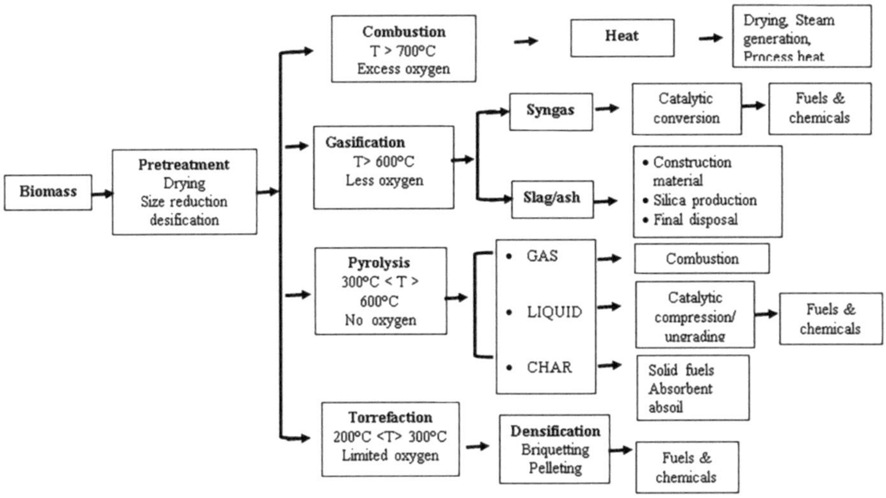

Figure 1.
Biomass thermal conversion processes.

lignocellulosic biomasses [56]. Torrefied biomass tends to be hydrophobic, brittle, and easily grindable, with significant energy and market potentials [42, 57, 58]. The process initiates at temperatures lower than 160°C with gradual removal of water, then advanced with hemicellulose degradation at temperatures between 180 and 270°C, accompanied by rapid biomass breakdown and release of moisture, volatile gases, and low-energy molecules that turn brown instead of yellow [59]. Torrefaction turns completely exothermic at 280°C, producing more gas and causing the synthesis of other heavy compounds and extraneous gases [56].

At temperatures above 300°C, but below 600°C in the absence of or limited air (oxygen), *pyrolysis* sets in as an irreversible thermal degradation of biomass [46, 53, 60]. The organic matter is broken down in an inert atmosphere without oxidizing agents, beginning at low temperatures of around 100°C with loss in moisture (dehydration), and at about 300°C, the biomass fractures and devolatilizes [61]. During pyrolysis, less oxygen is introduced as the temperature rises to 600°C, beyond this value, gasification results. The by-products of pyrolysis include gaseous products, ash, bio-oil, and other higher energy density materials [62]. Beyond pyrolysis temperature, combustion began. Combustion is a thermochemical process in fuel burning in air [63]. When fuel is heated in the air, combustion begins at the lowest temperature, which is the material ignition temperature, which varies for different materials. The shrinking core model of the burning of a briquette ball was used to explain the ignition temperature [64].

Fermentation: Fermentation is another pretreatment method requiring an application of solvents and/or chemicals such as water, acids, alkali, salts, and oxidants in breaking down lignocellulosic materials with the sole objective of lowering the volatile matter concentration while maintaining the structure of the cellulose and hemicellulose [65]. Several fermentation studies and procedures have been conducted to investigate comparative benefits in product energy contents, non-complex processes, better solubility, decreased volatility, higher octane number, and insoluble with water [66, 67]. Fermentation methods deployed in research studies include acid hydrolysis, alkali treatment and wet oxidation, organosolv pretreatment, and ionic and oxidizing processes have been developed [50]. Other fermentation approaches involved soaking materials in water for different lengths of soaking time [4] or in 5% NaOH for about 20 minutes and then transferring them into aqueous sodium hypochlorite (NaOCl) for 24 hours, then rinsing in water severally to improve materials transparency [68, 69]. This pretreatment is less severe than acid pretreatment due to the ease of experimentation performed under atmospheric conditions [70, 71].

In the fermentation process, biomass is soaked in a liquid medium for a while to lower the volatile matter concentration while preserving the cellulose and hemicellulose structure. Numerous existing fermentation techniques, each with unique contrasting benefits for energy contained product, ease of processing, and material applicability for further processing, improved miscibility with liquid fuels, higher octane ratings, and reduced volatility [67]. A literature search suggested torrefaction and fermentation as more attractive and less expensive processes for sawdust pretreatment for briquetting.

2. Briquetting technologies

Worldwide, a number of machines and techniques have been developed for briquetting on a range of different scales [72]. Several briquetting machines have been developed based on different operational mechanisms, including manual presses, screw presses, roller presses, pelletizing, and piston presses [4, 11, 73–75]. Mechanical piston press development started in Switzerland during World War II based on research works done in Germany in the 30's, which centered on Fred Housman and the Glomera presses [11, 43] and were typically common in Europe and the USA. The screw extrusion briquetting technology was invented in Japan in 1945 [11, 43] and preferred in the East. The integration of these two technologies is common in India and Bangladesh, however, it is relatively new to biofuel development technology in Africa compared with Asia, America, and Europe [2, 76].

The extent of development and utilization of briquette technology globally since 1992 has been significant and has seen the introduction of pillow-type coal briquettes by the Nepal Academy of Science and Technology (NAST) in collaboration with the Japan International Cooperation Agency (JICA) [77]. Subsequently, the Institute for Himalayan Conservation ("IHC"), in cooperation with NAST, introduced the roller presses in 1999. In 2000, NAST and the Asian Institute of Technology (AIT) introduced the screw-extruder briquetting technology to address screw-wearing problems [78–81]. Progressively from 2000, several research cooperations among international organizations and agencies across Asia had developed several briquetting machines comprehensively reported by Abdul Salam and Shobhakar [82]. Briquette extrusion had a long history of development and utilization, especially in industrial and domestic heating. In addition, several studies on different technologies, output capacity, shape and dimension, material used, and product characteristics were summarized by Kpalo et al. [2] and Bello et al. [3].

In Africa, briquetting technology is new [76], however, briquetting technology is gaining momentum, particularly as wood resources become scarcer and the price of regular charcoal increases in East Africa, where biomass dominates the domestic energy market [72, 77]. Nigeria's experience in briquetting technology is presently not quantitatively defined, as briquetting technology development efforts are limited to research engagements in research institutes, universities, and other tertiary institutions [78]. These efforts had produced several machines with different technologies for converting biomass into briquettes [82–84]. For instance, Sokoto Energy Research

Centre (SERC), in collaboration with the Usman Danfodio University, Sokoto developed a 13 kg/hr. capacity single-cylinder extruder to densify sawdust rice and millet husks. Obafemi Awolowo University, Ile-Ife Equally developed a 4 kW-powered screw press capable of producing 10 kg of briquettes/hr. with a calorific value of 17.8 MJ and 67.94% thermal efficiency [85]. Other successful briquetting technologies in Nigeria include the development of a 40 kg/hr. capacity piston press at the Center for Industrial Studies (CIS), Abubakar Tafawa Balewa University (ATBU), Bauchi, in collaboration with the Raw Material Research and Development Council (RMRDC), Abuja. Accelerated developmental efforts are ongoing at various research levels on medium/low-pressure densification with little attention on the development of high-pressure densification studies. Commercial briquetting installations with installed capacities as high as 2000 kg/hr. or more are limited in Nigeria, while there exist few small-scale industries producing and marketing briquette in Nigeria.

The viability of any briquetting technology depends on a number of key factors, such as energy inputs, technical complexities, potential markets for briquettes (i.e., costs as compared to other fuels), acceptability of briquettes by potential users, raw material availability, etc. To this end, any briquetting process that does not allow for real commercial standardization of the production process and sustainable investment in their production is considered not economically viable. Therefore, for economic evaluation, a review of the technoeconomic study of extrusion and compaction processes was undertaken using two machines (a screw extrusion briquetting technology and hydraulic piston briquetting technology) studied by the authors [3, 31].

2.1 Technologies reviewed

Screw-extruder machine: Briquetting technologies investigated include a screw extrusion briquetting technology [31] and hydraulic piston briquetting technology [3]. The extrusion machine (**Figure 2**) has three main functional units: the extrusion unit, the power transmission unit, and the power generation/electrical control systems. The extrusion unit comprises the extrusion barrel, shaft with screw assembly,

Figure 2.
Existing experimental screw press extruder.

universal connection, and die that make up the extrusion unit. The barrel is a long cylindrical pipe with sufficient thickness (to sustain severe compressive stresses) that is supported by two L-shaped mount brackets attached to the frame with four 21 mm bolts and nuts. Toward the bearing end is a fixed hopper for material infeed into the machine, while at the other end is a welded circular bracket for the die attachment. The screw shaft is a long-tapered shaft with a machined single-flight screw conveyor that extends from the end of the extrusion section into the die ring. A massive trust bearing attached to one end of the cylinder serves as the shaft's support (cantilever). A ring flange that is welded to the die connects a short, cylindrical metal pipe with an internal conic taper at the barrel end that is bored into a uniformly cylindrical hole at the other end, known as the die [31].

The power transmission unit comprises a speed reducer assembly, belt, double-grooved pulley, and universal coupling. The extruder receives power from a 3-phase, delta-connection induction motor. The instrument control system comprises short-circuit protection, a contactor, a temperature controller, a temperature sensor, and a heating element connection. Short-circuit protection shuts off the heater's power supply above a predetermined temperature. It then trips when the temperature falls below that threshold, raising the raw material temperature for improved densification. There are two heaters with dimensions of length 100 mm and diameter 90 mm and rated power outputs of 1300 and 1500 W, respectively [31]. To ensure that the temperature requirements for extrusion are met, a J-type thermocouple sensor and a PID-JTC 9036.3 A, 415 V temperature device were installed to regulate the heater temperature. The power generation unit consists of the induction motor, which supplies power to drive the extrusion shaft and provide power for short-circuit protection and direct induction heating via two mica-type band heaters attached to the die. The machine functional requirement, component description, technical specifications, and quantity for individual machine components are presented in **Table 2** below.

S/N	Item	Description/specification
1	Induction motor	20 HP (15 kW) @ 1460 rpm, 380 volts/50 Hz, 3-phase Δ/λ connection
2	V-Belts	B-90
3	Pulleys	12.5 cm dia. (driver), 47.0 cm dia. (driven)
4	Reduction gear assembly	Rear-axle reduction gearbox
5	Bearings	N 6312, N 6311
6	Transmission shaft	3″ dia.
7	Die	9.7 cm dia., 30 cm long (MS)
8	Screws	2 ¼“dia., 45 cm long (MS)
9	Bushing	Outside dia: 73 mm, inside dia: 59 mm length: 32 mm
10	Frame	Length = 256 mm, width = 88 mm, and height = 98 mm
11	Control panel	Electrical distribution board
12	Extrusion barrel	Cylindrical pipe diameter

Source: Bello and Onilude [31].

Table 2.
Technical details of the experimental machine components.

Figure 3.
The developed hydraulic briquetting machine and production. A. Material loading, B. Material compression, C. Removing mold.

Hydraulic Piston Press [HPP]: The hydraulic piston briquetting press comprises four pistons equipped with two balances symmetrically positioned alongside a centrally positioned hydraulic jack to measure the load exerted by spring balances on each piston (**Figure 3A**). The frame structure, which supports the entire machine, was made of mild steel (MS) angle iron that was 50 mm wide and 3.0 mm thick. The hydraulic jack is a 5-tonne machine that provides the necessary force to compress the biomass material mix in the cylindrical mold cups. The jack was fastened to the frame with a 24 mm bolt and the nut welded to the extension bar above, and it was secured to the pressure plate with four 8 mm bolts and nuts that hold the four-piston plates beneath [3].

To measure the load/force given to the loaded biomass under compression, two symmetrically mounted dynamically balanced helical extension spring balances were placed at either side of the jack (**Figure 3B**). The springs expanded with the movement of the pressure plate through the action of the hydraulic jack at compression. Each piston was dynamically positioned and fastened on a 3 mm thick flat base plate, with each fused to a 2 mm thick piston-base plate at one end and a 20 mm diameter pipe at the other end. The pistons transfer the compressive pressure to the material charge within the four cylindrical mold cups positioned on a 420 x 420 mm hardwood pressboard. The spacers made of plastic are available to produce several briquettes in a cylinder mold. After compression, a detachable rod inserts with a diameter of 20 mm and a length of 240 mm was placed into the piston pipe to create the briquette's center hole (**Figure 3C**).

Table 3 shows the summary of the designed parameters for the Hydraulic Piston Press [HPP]. The Hydraulic Piston Press [HPP] performance was evaluated based on briquette quality and physical and mechanical characteristics. The machine produced durable and stable hollow briquettes at an operating pressure of 167 kPa.

2.2 Operational variables

The selection of briquetting technology is dependent on desired performance characteristics such as heat source, pressure application, material type and

S/No	Parameters	Values
1	Load applied by jack	$1.99\,MN$
2	Maximum torque transmitted	$79{,}738.82\,Nm$
3	Cylinder dimensions	Height = 151.36 mm, diameters = 78.50 mm
4	Height of briquettes produced	50–60 mm
5	Buckling load	2.932 MN
6	The area of each piston plate	$0.071\,m^2$
7	Briquetting pressure	1.66 kPa
8	Circumferential stress on the cylinder	$3.735\,N / m^2$
9	Longitudinal stress on the cylinder	$1.87\,N / m^2$
10	Compressive stress on cylinder hanger	$6.35\,MNm^{-2}$
11	Power requirement	$26.49\,Nm$ / sec
12	Spring constant	9.81 kN / m^2

Source: Bello et al. [3].

Table 3.
Designed values of hydraulic briquetting machine.

characteristics, power consumption, and energy value, among others, with the major determinants of the choice of technology being the biomass material (material-specific) characteristics, non-material-specific variables [86], product quality, and production costs [87]. Process variables dominant in briquette production include the operating temperature, pressure, and dwell time. These factors have a prominent influence on the briquette's physical properties, densities (compressed and relaxed), and durability.

In the screw extrusion, heat is a principal requirement in successful operation. The secondary source of material preheating was the inter-particle interactions during high-pressure compaction, while the primary source was using two external mica-heating bands rated at 1300 and 1500 W. The use of heating elements across the entire die barrel surface allows excellent heating over a wide range of temperatures up to 350°C [88]. The use of heating the die with thermal band heaters has advantages including high-watt density and temperature capabilities, good efficiency, a good lifetime, and a low cost over other types of heaters. The total heat energy (kWh) required in extrusion depends on the start-up heat and additional heat required to maintain the process at the desired temperature. The start-up power and required operating power (kW) are expressed as heat energy (kWh) per unit start-up time. Power losses during start-up time are zero and may rise to as much as 100% at peak process temperature due to surface heating, conduction, and radiation. In addition, because of the variability in material characteristics, the amount of heat required varied and, subsequently, temperature changed. However, differences in material types and characteristics under varying process conditions produced different qualities of the final products [13, 89].

Hydraulic piston briquetting employed compressive force in the densification process without heat requirement. Two operating conditions considered during briquette production include residence time and operating pressure. A residence time of 5 minutes was allowed for the compressed material to remain under compressive load before ejection [3]. The dwell time does not significantly affect briquette durability but has a significant effect on relaxation properties. An operating pressure of 169.80 kPa and an axial load of 45 kg were considered suitable for briquette production.

2.3 Technical analysis and operations

Extrusion is a highly technical process requiring accuracy, precision, and correct operating conditions. Advances in extrusion technology and more rapid product changes demand an adaptable high extrusion performance with outstanding throughput without changing screw or barrel design. Screw and barrel designs adaptable to a gradual increase in pressure along the extruder and low friction in the feed section are required in extruder design technology for quality briquette production at the lowest costs required. The conical screw shaft within the enclosed barrel protruding into the die requires precision to synchronize conveyed material into uniform regular or irregular cross-sections identical to cylindrical bars or hollow tubes with that of the opening [25]. Misalignment caused discontinuity in material flow from the barrel into the die at the transition zone, hence causing deformation and leading to failed extrusion. Machine abnormalities that could cause poor performance are presented in **Table 3**, while the summary of the observable performance characteristics of the extruder during briquette production is presented in **Tables 4** and **5**.

Several research efforts have been made to optimize screw design and fabrication methods [90]. These efforts require specifying the extruder screw geometry to offer a 90% minimum reduction volume or 10% relative final volume and applying the principles of mechanics to determine the applied forces and power consumption requirement to operate the machine. In addition, specifying the geometry of the die, size (diameter), as well as the temperature of the heating ring. The specific output capacity of the extruder is dependent on the barrel and screw geometry derivatives.

Technical considerations for the design of Hydraulic Piston Press [HPP] are simple compared with those of a screw press. The considerations for the design of

Components	Design errors	Suggested corrections
Screw and shaft	Irregular profiles at conical section tip, high taper angle (15°) relative to the shaft length.	Redesign the shaft, screw length, and reduce the taper angle (≈3.2°).
Screw-die clearance	Too small (≈3 ± 0.02 mm).	Increase die clearance by (≈5 ± 0.02 mm).
Die cone	Spotted weld within die cone.	Redesign.
Shaft length to taper ratio	High (4:1).	Reduce the ratio (3:1).
Barrel and die inner surface	Spot weld projections within the barrel and die.	Remove the projections to have a smooth surface.
	Uneven barrel and die surfaces.	Machine the surfaces.
Barrel length	Short barrel length between transport and compression zones.	Redesign barrel.
Die length	Short die length.	Increase the length.
Heating element	Low power rating (1300 W).	Increase power rating.

Table 4.
Results of observation on existing machine performance.

Machine factors	Extruder performances
Material transport	Within the transition zone in the barrel.
Material compression	Occurred within the compression zone.
Dwell time	3–4 minutes approx.
Extrusion characteristics	Well-formed materials are extruded.
Product defects	Products showed hairpin cracks but were well-formed.
Briquette quality	Well-formed and strong briquettes with brownish-char coloration.
Product durability	Extruded products are durable, stable, and store longer.

Table 5.
Summary of performance characteristics of extruder.

Parameters	Piston press	Screw press
moisture content of biomass	10–15%	8–9%
Wear of contact press	Low in case of ram and die	High
Output from the machine	In strokes	Continuous
Power consumption	30 kWh/tonne	60 kWh/tonne
Density of briquettes	1–1.2 gms/cm^3	1–1.4 gms/cm^3
Maintenance	High	Low
Combustion performance	Inclined to crumble on grate, smoky	Burns well with minimal smoke
Carbonization to charcoal	Not possible	Possible
Suitability in gasifiers	Not suitable	Suitable
Briquette homogeneity	Non-homogeneous	Homogeneous

Table 6.
Comparison between a screw press and a piston press.

the hydraulic briquetting machine include that the cylinder wall thickness selected should be able to withstand pressures due to the compressive action of the piston on materials, pressure generated by the jack was transmitted to the cylinder wall via the piston surface, and the torque transmitted by the jack to the pistons should be sustained without failing. The machine performance evaluation showed that Hydraulic Briquetting Press [HBP] produced durable and stable hollow briquettes and had an output capacity of 96 briquettes/hour. The machine was simple in operation, suitable for low-pressure densification, and inexpensive to produce compared to screw press operation. **Table 6** shows the comparison between a screw press and a piston press.

Process efficiency: Improving the process efficiency in briquetting is one of the major objectives for machine development. Technical characteristics of a screw press machine that could affect the process efficiency are associated with shaft and screw geometry, feeder capacity, and extrusion horsepower requirements. Another critical factor that adversely limits the machine productivity and final product quality is a defective die. A reduction in die-material frictional relationship, die taper angle, and material resident time was able to reduce such defects and improve the grain structure boundaries around the tolerance or electroplated die contact surface. In the screw extruder, maximizing efficiency requires production of good-quality products and the ability to process a broad range of raw materials with the same screw, which contribute to the economy of a machine. Machine production soundness, surface finishing, poor material selection, and cost considerations all affect machine performance. The effect of the biomass material factors, its particle moisture, on the capacity of briquetting machine efficiency is used to evaluate the machine performance and product characteristics.

Extruder operations are expected to be optimized for maximum performance; however, existing installations in Nigeria are either moribund or operating below the rated capacities, producing lower-quality products, reducing the profitability of the plant [91]. Operating at the maximum rate will minimize the fixed costs of the product by spreading these costs over more product. High economic efficiency is achieved by low specific investment costs relating to throughput and low operating costs [92]. Other important aspects include low energy consumption and low maintenance costs.

2.4 Power consumption pattern

Power consumption is a critical factor in the selection of briquetting technology, viability of any briquetting operation, and a major determinant of product quality as well as the production costs. The power consumptions for hydraulic Piston presses are lower compared to those of the screw presses, which require electricity to drive the shaft and heat up the mica band. The total horsepower requirement in screw presses depends on accurate installation (precision), constant feed rate, flight width factor, which has a relative contribution to power consumption. Feedstock pretreatment is another factor that influences the power consumption pattern. Feedstock pre-heating could reduce power consumption to 30%, lower die temperature, and reduce screw wear, thereby increasing screw lifespan up to 44 hrs during briquetting [26, 30]. Material build-up within the extruder could also led to high power consumption with material sticking around the root of the screw flight, and subsequent sticking around the entire shaft due to screw surface roughness and irregular shaft profile at the periphery of the conic section. The power consumption pattern in HPP is low compared with screw presses due to the wet-material charge fed to the machine; however, screw press has better production performance and product quality at the expense of high power consumption. Considering the energy values of products produced, there are strength limitations in compression press products at the expense of high machine output but high operational cost to meet the demands for solid fuel in developing countries [91, 93].

Sieve mesh	Description	Particle size	Particle length	Mean particle length
#8	Oversize (OS)	2.38 mm	< 10.67 mm	8.57 mm, Standard deviation at α = 95%, 4.594
#12	Coarse pin size (CPS)	1.70 mm	< 9.98 mm	
#20	Coarse pin particle (CPP)	0.85 mm	< 8.64 mm	
#35	Fine pin particles (FPP)	0.50 mm	< 4.5 mm	
>#40	Fine pin size (FPS)	0.40 mm	< 2.07 mm	

Source: Bello et al. [3].

Table 7.
Particle-size distribution of G. arborea *sawdust.*

2.5 Feedstock materials and characteristics

The dominant lignocellulosic material used is sawdust from the Gmelina arborea species for reasons of volumetric availability in large quantities, varied utilization in waste-to-wealth conversion processes, and utilization in combating environmental menace [32, 55, 76]. Mill wastes were sourced randomly from sawmills and other mills, which were sorted for foreign matters and prepared as received, requiring no size reduction. The materials were sundried to moisture contents favorable for the production of stable briquettes (8–12%) dry basis and characterized [94]. Material characterization parameters determined include the particle-size distribution, particle density, geometric mean sizes, and particle moisture [30, 95]. The summary of particle-size distribution, particle length, and mean particle lengths determined are shown in **Table** 7 [3].

3. Economic analysis of briquetting processes

Economic analysis of briquetting process centers around assessment of the economic feasibility of the production and utilization of briquettes and consequently the market perspectives, keeping in focus the value additions compared with other studies. The key focus is the return on investment (RoI) and the value addition when compared to raw materials. Commercially developed markets and pricing structures for briquettes are not readily available with assumptions of premium payment for potentially superior storable products similar to those made from coal [96]. Several economic assessment models developed are available to evaluate the cost of material pretreatment and briquetting [97]. The assessment model includes an analysis of the costs of biomass treatment, electricity, labor, investment, and transportation [96].

Cost variables	Piston press (dry ton/yr), NGN	Screw press (dry ton/yr), NGN
Capital cost of machine	150,000	500,000
Cost of machine installation	80,000	150,000
Cost of biomass preparation/ton	20,000	20,000
Biomass transportation cost/ton	55,000	55,000
Cost of binder preparation/ton	30,000	—
Labor costs for production/ton	25,000	25,000
Unit production cost	3000	3000
Operating cost	15,000	15,000
Cost of energy input	30,000	80,000/kWh
Manpower cost	15,000	15,000
Miscellaneous cost	10,000	35,000
Total production cost	433,000	898,000
Cost of final product	200/kg	450/kg

Table 8.
Estimated economic evaluation of screw press and piston press.

These analyses reported a production cost savings of over 3% above pelletization with considerable transport and end user savings [96].

Economic viability: The scale of preference of briquetting technology is tied to the economic viability of the material availability, technology deployed, and energy consumption in the production of unit products [98]. The economic viability of any system assesses the operational profitability of such system and compares the analyzed procedure with other more intensive productive systems. Different methods have been used to evaluate economic viability, ranging from standardizing methods and operational productivities to evaluate the operational unit costs [99, 100], and profitability evaluated through the estimation of Net Present Values (NPV) and Internal Rates of Return (IRR) [101]. In order to provide an economic analysis of the briquetting technologies and the main products from biomass (i.e., electricity and biofuels), an assessment of the production cost, conversion efficiency, and size of the processing is required.

Production costs: Economic cost variables for the production of briquettes include the cost of equipment and installation, operating costs, and equipment maintenance costs. The capital costs include the cost of pre-processing equipment, material handling equipment, and the briquetting press, depending on the type of raw material. The installation cost includes the cost of equipment installation at the site. The operating cost includes the cost of various inputs, such as raw material, oil lubricant, and electricity, as well as the manpower required to operate the unit. The cost of energy was identified as a major barrier to achieving sustainability for biomass briquetting technology. Operating the screw press is not economically viable because of the high electricity tariffs, (NGN80, 000/kWh) compared with (NGN30, 000/kWh) for HPP, poor electricity supplies challenges in Nigeria and high precisions required in component manufacturing and installation.

Table 8 shows the comparison of estimated economic evaluation of cost component considerations for screw press and piston press technologies, which include the cost of machine, installation, biomass processing, binder cost, energy input (NGN/kWh), and the cost of the final product (NGN/kg) [102, 103]. The use of alternative energy facilities such as diesel engines as sources of power and die heating has reduced the cost of briquette production to 50%.

4. Conclusions

The technoeconomic analysis of screw press and piston press briquetting technology has been reviewed from the perspective of technical production and technology, which revolves around the technical evaluations, based on biomass material and briquetting technologies. The biomass technical variables reviewed include material availability, material pretreatments, and characteristics. The reviewed technical variables around the briquetting technologies include operational variables, machine performance variables, process efficiency, and power consumption patterns. The economic analysis reviewed consists of the cost of biomass, briquetting processes, and market values of final products. The economic feasibility of sustainable production, keeping return on investment (RoI) and value additions in focus.

The economic analysis of the briquetting process revealed considerable cost and end-user savings in HPP briquetting over SP at the expense of energy value of final products. The economic viability of each process based on different factors of production depended on the material availability, technology deployed, and energy consumption in the production of unit products. The energy cost was a major barrier to achieving sustainability for biomass briquetting with screw press technology. Operating the screw press is not economically viable because of the high electricity tariffs, poor electricity supplies, and high level of machine precisions required. However, the use of alternative energy facilities such as diesel engines as a source of power and die heating has reduced the cost of briquette production to 50%.

The technoeconomic benefits of screw and piston press technologies, product characteristics, and economic analysis showed that the screw press significantly improved the physical quality and combustion performance of products compared with the piston press. The screw press products have better storability compared to piston press; however, the cost of machine acquisition, installation, and operation of piston press is much lower than screw press. The quality of extruded products is higher than the piston press, and its high-density characteristics translate to an economic advantage over the high cost of production. A critical technical challenge in the development of the screw press is that of alignment of the extrusion head with the die. This challenge and high cost of electricity tariff discouraged the availability of small-scale screw presses.

Conflict of interest

The authors have no conflict of interest in this work.

Author details

Raphael Segun Bello[1*], Abel Olajide Olorunnisola[2], Temidayo Emmanuel Omoniyi[2] and Musiliu Ademuiwa Onilude[2]

1 Department of Agricultural and Bioenvironmental Engineering Technology, Federal College of Agriculture Ishiagu, Nigeria

2 Department of Wood Products Engineering, University of Ibadan, Nigeria

*Address all correspondence to: segemi2002@fcaishiagu.edu.ng

References

[1] Sansaniwal SK, Pal K, Rosen MA, Tyagi SK. Recent advances in the development of biomass gasification technology: A comprehensive review. Renewable and Sustainable Energy Reviews. 2017;**72**:363-384

[2] Kpalo SY, Zainuddin MF, Manaf LA, Roslan AM. A review of technical and economic aspects of biomass briquetting. Sustainability. 2020;**12**:4609. DOI: 10.3390/su12114609 www.mdpi.com/journal/sustainability

[3] Bello RS, Olorunnisola AO, Omoniyi TE, Onilude MA. Development of multiple-piston hydraulic briquetting press [HBP] and characterization of briquettes produced. Trends in Agriculture Sciences. 2023a;**2**(2):169-188. DOI: 10.17311/tas.2023.169.188

[4] Bello RS, Olorunnisola AO, Omoniyi TE, Onilude MA. A review of technoeconomic benefits of torrefaction pretreatment technology and application in torrefying sawdust. Sustaining Economies. 2024;**2**(2):104. DOI: 10.62617/se.v2i2.104

[5] Hussain J, Hassan S. Global energy transition and the role of energy mix in creating energy crisis in Pakistan. Pakistan Journal of Humanities and Social Sciences. 2019;**7**:219-232

[6] Olatayo JS, Orisaleye JI, Pecenka R. An assessment of potential resources for biomass energy in Nigeria. Resources. 2020;**9**:92. DOI: 10.3390/resources9080092

[7] Tokan A, Sambo AS, Jatau JS, Kyauta EE. Effects of particle size on the thermal properties of sawdust, corncobs and prosopis Africana charcoal briquettes. American Journal of Engineering Research. 2014;**3**(8):369-374. Available from: http://www.ajer.org/papers/v3(8)/ZM03803690374.pdf

[8] Kiss I, Alexa V. Short introspections regarding the sawdust briquetting as sustainable solution for the environment. Analecta. 2015;**8**(2):72-79. ISSN 2064-7964

[9] Antonio T. A review on biomass: Importance, chemistry, classification, and conversion. Biofuel Research Journal. 2019;**22**:962-979. DOI: 10.18331/BRJ2019.6.2.3

[10] Kaliyan N, Morey RV. Densification Characteristics of Corn Cob. St. Joseph Michigan: American Society of Agricultural and Biological Engineers; 2008

[11] Oladeji JT. Theoretical aspects of biomass briquetting: A review study. Journal of Energy Technologies and Policy. 2015;**5**(3):72-81

[12] Granada E, López GLM, Míguez JL, Moran J. Fuel lignocellulosic briquettes, die design and products study. Renewable Energy. 2002;**27**:561-573

[13] Križan P, Matúš M, Šooš L, Beniak J. Behaviour of beech sawdust during densification into a solid. Biofuel Energies. 2015;**8**:6382-6398. DOI: 10.3390/en8076382. ISSN 1996-1073. Available from: www.mdpi.com/journal/energies

[14] International Energy Agency (IEA). Africa Key Energy Statistics. 2002. Available from: http://www.iea.org

[15] Kipyego EK. Optimization of biomass use and performance as an industrial fuel: Experimental study [MSc

thesis]. Kenya: Jomo Kenyatta University of Agriculture and Technology; 2009. pp. 4-6

[16] Sen R, Wiwatpanyaporn S, Annachhatre AP. Influence of binders on physical properties of fuel briquettes produced from cassava rhizome waste. International Journal of Environment and Waste Management. 2016;**17**(2):158-175. DOI: 10.1504/IJEWM.2016.076753

[17] Hamid MF, Idroas MY, Ishak MZ, Zainal Alauddin ZA, Miskam MA, Abdullah MK. An experimental study of briquetting process of Torrefied rubber seed kernel and palm oil shell. BioMed Research International. 2016;**2016**(1):1679734

[18] Srivastava NSL, Narnaware SL, Makwana JP, Singh SN, Vahora S. Investigating the energy use of vegetable market waste by briquetting. Renewable Energy. 2014;**68**:270-275. DOI: 10.1016/j.renene.2014.01.047

[19] Gill N, Dogra R, Dogra B. Influence of moisture content, particle size, and binder ratio on quality and economics of Rice straw briquettes. Bioenergy Research. 2018;**11**:54-68

[20] Moreno AI, Font R, Conesa JA. Physical and chemical evaluation of furniture waste briquettes. Waste Management. 2016;**49**:245-252. DOI: 10.1016/j.wasman.2016.01.048

[21] Manyuchi MM, Mbohwa C, Muzenda E. Value addition of coal fines and sawdust to briquettes using molasses as a binder. South African Journal of Chemical Engineering. 2018;**26**(1):70-73. DOI: 10.1016/j.sajce.2018.09.004

[22] Thoreson CP, Webster KE, Darr MJ, Kapler EJ. Investigation of process variables in the densification of corn Stover briquettes. Energies. 2014;7(6):4019-4032. DOI: 10.3390/en7064019

[23] Borowski G, Stępniewski W, Wójcik-Oliveira K. Effect of starch binder on charcoal briquette properties. International Agrophysics. 2017;**31**(4):571. DOI: 10.1515/intag-2016-0077

[24] Dinesha P, Kumar S, Rosen MA. Biomass briquettes as an alternative fuel: A comprehensive review. Energy Technology. 2018;**1**:1-21

[25] Mithun KM. Three-dimensional analysis of extrusion through taper die [MSc thesis in mechanical engineering]. National Institute of Technology Rourkela; 2009

[26] Adegoke CO, Alatise MO, Mohammed TI. Development of a Manually Operated Hydraulic Briquetting Machine. Personal Communication. Akure, Nigeria: Federal University of Technology; 1999

[27] Bhattacharya SC, Augustus LM, Rahman MM. A study on improved biomass briquetting. Energy for Sustainable Development. 2002;**6**(2):67-71

[28] Mani S, Tabil LG, Sokhansanj S. An overview of compaction of biomass grinds. Powder Handling and Process. 2003;**15**(3):160-168

[29] Olorunnisola AO. Production of fuel briquettes from waste paper and coconut husk admixtures. Agricultural Engineering International. 2007;**1X**:1-11. article EE06066. Available from: https://hdl.handle.net/1813/10628

[30] Adapa PK, Tabil LG, Schoenau GJ. Compression characteristics of selected ground agricultural biomass. Agricultural Engineering International:

The CIGR Ejournal. Manuscript 1347. 2009;**XI**:1-19

[31] Bello RS, Onilude MA. Effects of critical extrusion factors on quality of high-density briquettes produced from sawdust admixture. In: Materials Today: Proceedings. Vol. 38. Elsevier; 2021. pp. 949-957. DOI: 10.1016/j.matpr.2020.05.468

[32] Okoroigwe E, Li ZL, Stuecken T, Saffron C, Onyegegbu S. Pyrolysis of *Gmelina arborea* wood for bio-oil/bio-char production: Physical and chemical characterisation of products. Journal of Applied Sciences. 2012;**12**(4):369-374. DOI: 10.3923/jas.2012.369.374

[33] Ismail S, Ojolo S, Orisaleye J, Olusegun F. Design of an o_ce table solar-DC powered fan. Journal of Emerging Trends in Engineering and Applied Sciences. 2014;**5**:1-5

[34] Orisaleye JI, Ogbonnaya M, Ogundare AA, Ismail SO. Development and performance evaluation of a natural draft mixed-type solar dryer for agricultural products. Journal of Science and Technology. 2018;**10**:18-24

[35] Keles S, Bilgen S, Kaygusuz K. Biomass energy source in developing countries. Journal of Engineering and Applied Science. 2017;**6**:566-576

[36] Orisaleye J, Ismail S, Ogbonnaya M, Ogundare A. Development and performance evaluation of a solar water still. The Acta Technica Corviniensis—Bulletin of Engineering. 2018a;**11**:91-96

[37] Osiolo HH. Green Energy and its Impact on Employment and Economic Growth; Paper 19. Accra Ghana: United Nations University Institute for Natural Resources in Africa (UNU-INRA); 2016

[38] Kumar A, Kumar N, Baredar P, Shukla A. A review on biomass energy resources, potential, conversion and policy in India. Renewable and Sustainable Energy Reviews. 2015;**45**:530-539

[39] Habert G, Bouzidi Y, Chen C, Jullien A. Development of a depletion indicator for natural resources used in concrete. Resources, Conservation and Recycling. 2010;**54**(6):364-376

[40] Colin S, Schuck S, Sims R, Bland J, Marino B, Borowitzka M, et al. Bioenergy in Australia. In Status and Opportunities. Vol. 15. St. Leonards: Bioenergy Australia (Forum) Limited; 2012

[41] Matali S, Rahman NA, Idris SS, Yaacob N, Alias AB. Lignocellulosic biomass solid fuel properties enhancement via torrefaction. Procedia Engineering. 2016;**148**:671-678. DOI: 10.1016/j.proeng.2016.06.550

[42] Nhuchhen DR. Studies on Advanced Means of Biomass Torrefaction. 2016. Available from: http://hdl.handle.net/10222/71402

[43] Grover PD, Mishra SK. Biomass briquetting technology and practices. Food and Agriculture Organization (FAO). UN, Document. 1996;**46**:1-48

[44] Kumar SR, Sarkar A, Chakraborty JP. Effect of torrefaction on the physicochemical properties of pigeon pea stalk (*Cajanus cajan*) and estimation of kinetic parameters. Renewable Energy. 2019;**138**:805-819. DOI: 10.1016/j.renene.2019.02.022

[45] Rosillo-Calle F. A review of biomass energy–shortcomings and concerns. Journal of Chemical Technology and Biotechnology. 2016;**91**(7):1933-1945. DOI: 10.1002/jctb.4918

[46] Kaniapan S, Pasupuleti J, Patma Nesan K, Abubackar HN, Umar HA,

Oladosu TL, et al. A review of the sustainable utilization of Rice residues for bioenergy conversion using different valorization techniques, their challenges, and techno-economic assessment. International Journal of Environmental Research and Public Health. 2022;**19**(6):3427. DOI: 10.3390/ijerph19063427

[47] Mohapatra S, Mishra C, Behera SS, Thatoi H. Application of pretreatment, fermentation and molecular techniques for enhancing bioethanol production from grass biomass–A review. Renewable and Sustainable Energy Reviews. 2017;**78**:1007-1032. DOI: 10.1016/j.rser.2017.05.026

[48] Akhtar N, Gupta K, Goyal D, Goyal A. Recent advances in pretreatment technologies for efficient hydrolysis of lignocellulosic biomass. Environmental Progress and Sustainable Energy. 2016;**35**(2):489-511. DOI: 10.1002/ep.12257

[49] Brunerová A, Roubík H, Brožek M, Herák D, Šleger V, Mazancová J. Potential of tropical fruit waste biomass for production of bio-briquette fuel: Using Indonesia as an example. Energies. 2017;**10**(12):2119. DOI: 10.3390/en10122119

[50] Bensah EC, Mensah M. Chemical pretreatment methods for the production of cellulosic ethanol: Technologies and innovations. International Journal of Chemical Engineering. 2013;**2013**(1):21. DOI: 10.1155/2013/719607

[51] Wagner AO, Schwarzenauer T, Illmer P. Improvement of methane generation capacity by aerobic pretreatment of organic waste with a cellulolytic *Trichoderma viride* culture. Journal of Environmental Management. 2013;**129**:357-360. DOI: 10.1016/j.jenvman.2013.07.030

[52] Shah A, Darr M, Anex R, Khanal S, Maski D, Medic D. Techno-economic analysis of a production-scale torrefaction system for cellulosic biomass upgrading. Biofuels, Bioproducts, and Biorefining. 2012;**5**:45-57. DOI: 10.1002/bbb.336

[53] Maguyon-Detras MC, Migo MVP, Van Hung N, Gummert M. Thermochemical conversion of Rice straw. In: Gummert M, Hung N, Chivenge P, Douthwaite B, editors. Sustainable Rice Straw Management. Cham: Springer; 2020. DOI: 10.1007/978-3-030-32373-8_4

[54] Vaish S, Sharma NK, Kaur G. A review on various types of densification/briquetting technologies of biomass residues. In: IOP Conference Series: Materials Science and Engineering. Vol. 1228, No. 1. IOP Publishing; 2022. p. 012019. DOI: 10.1088/1757-899X/1228/1/012019

[55] Bello RS, Olorunnisola AO, Omoniyi TE, Onilude MA. Effects of binder concentrations and soaking time on combustion characteristics of briquettes produced from fermented *Gmelina Arborea (Roxb)* sawdust and used print paper. Bioenergy Studies. 2022;**2**:031-042. DOI: 10.51606/bes.2022.9

[56] Tumuluru JS, Sokhansanj S, Wright CT. Biomass Torrefaction Process Review and Moving Bed Torrefaction System Model Development (No. INL/EXT-10-19569). Idaho Falls, ID (United States): Idaho National Lab (INL); 2010. DOI: 10.2172/1042391

[57] Adnan MA, Mohd Fuad MAH, Hasan MF. Oxidative torrefaction for pulverized palm biomass using air. Jurnal Teknologi. 2017;**79**:7-14. DOI: 10.11113/jt.v79.12259

[58] Basu P. Biomass Gasification, Pyrolysis and Torrefaction: Practical Design and Theory. 2nd ed. Dalhousee University, Academic press; 2018. pp. 479-488. DOI: 10.1016/C2016-0-04056-1

[59] Zanzi R, Ferro DT, Torres A, Soler PB, Bjornbom E. Biomass torrefaction. In: The 6th Asia-Pacific International Symposium on Combustion and Energy Utilization. Kuala Lumpur: The Hong Kong Polytechnic University; May 20-22, 2002

[60] Pollex A, Ortwein A, Kaltschmitt M. Thermo-chemical conversion of solid biofuels. Biomass Conversion and Biorefinery. 2012;**2**(1):21-39. DOI: 10.1007/s13399-011-0025-z

[61] Weldekidan Pradhan P, Mahajani SM, Arora A. Production and utilization of fuel pellets from biomass: A review. Fuel Processing Technology. 2018;**181**:215-232. Available from: https://link.springer.com/article/10.1007/s13399-011-0025-z

[62] Uddin MN, Techato K, Taweekun J, Rahman MM, Rasul MG, Mahlia TMI, et al. An overview of recent developments in biomass pyrolysis technologies. Energies. 2018;**11**(11):3115

[63] Huttunen S, Lampinen A. Bioenergy Technology Evaluation and Potential in Costa Rica (No. 81). University of Jyväskylä; 2005. Available from: http://urn.fi/URN:ISBN:951-39-2154-9

[64] Pandey S, Dhakal RP. Pine needle briquettes: A renewable source of energy. International Journal of Energy Research. 2013;**3**(3):254-260

[65] Amin FR, Khalid H, Zhang H, Zhang R, Liu G, Chen C. Pretreatment methods of lignocellulosic biomass for anaerobic digestion. AMB Express. 2017;**7**(1):1-12. DOI: 10.1186/s13568-017-0375-4

[66] Kumar AK, Sharma S. Recent updates on different methods of pretreatment of lignocellulosic feedstocks: A review. Bioresources and Bioprocessing. 2017;**4**(1):1-19

[67] Shi S, Yue C, Wang L, Sun X, Wang Q. A bibliometric analysis of anaerobic digestion for butanol production research trends. Procedia Environmental Sciences. 2012;**16**:152-158. DOI: 10.1016/j.proenv.2012.10.021

[68] Jayeola AA, Aworinde DO, Folorunso AE. Use of wood characters in the identification of selected timber species in Nigeria. Notulae Botanicae Horti Agrobotanici Cluj-Napoca. 2009;**37**(2):28-32. Available from: https://scholar.oauife.edu.ng/sites/default/files/afolorun/files/jayeola_et_al._2009.pdf

[69] Olusegun-Awosika BM, Adegunloye DV. Potential of Brachystegia Nigeria sawdust as fermentation medium for biobutanol production: A preliminary study. Journal of Energy Technologies and Policy. 2020;**10**(7):6-15. DOI: 10.7176/JETP/10-7-02

[70] Menon V, Rao M. Trends in bioconversion of lignocellulose: Biofuels, platform chemicals and biorefinery concept. Progress in Energy and Combustion Science. 2012;**38**(4):522-550. DOI: 10.1016/j.pecs.2012.02.002

[71] Yoo HY, Kim SB, Choi HS, Kim K, Park C, Kim S. Optimization of sodium hydroxide pretreatment of canola agricultural residues for fermentable sugar production using statistical method. IPCBEE. 2012;**28**:175-179. Available from: https://environment.scientific-journal.com/articles/5/28.pdf

[72] Lamido SI, Lawal M, Salami H. Briquetting business in Nigeria: A

solution to unemployment. International Journal of Engineering Development and Research. 2018;**6**(4):101-106

[73] Chen L, Xing L, Han L. Renewable energy from agro-residues in China: Solid biofuels and biomass briquetting technology. Renewable and Sustainable Energy Reviews. 2009;**13**:2689-2695

[74] Tumuluru SJ, Wright CT, Hess JR, Kenney KL. A review of biomass densification systems to develop uniform biomass commodities for bioenergy application. Biofuels, Bioproducts and Biorefining. 2011;**5**:683-707

[75] Kaur A, Roy M, Kundu K. Densification of biomass by briquetting: A review. The International Journal of Recent Scientific Research. 2017;**8**:20561-20568

[76] Yoshida K, Singh RM. Summary Report - Biomass Sector. Kathmandu, Nepal: RONAST; 1995

[77] Bhattacharya SC, Kumar S. Renewable Energy Technology in Asia: A Regional Research and Dissemination Programme Technology Packages: Screw-Press Briquetting Machines and Briquette-Fired Stoves. 2005. Available from: www.serd.ait.ac.th/reric/

[78] Poudyal R, Regmi S. Review on Future of Biomass Energy: A Case of Nepal. Engineering for Environment. Proc. of the 8th SONEUK Conference, Pub. Society of Nepalese Engineers in UK (SONEUK). 2023. p. 63

[79] Obi OF, Akubuo CO, Okonkwo WI. Development of an appropriate briquetting machine for use in rural communities. International Journal of Engineering and Advanced Technology. 2013;**2**(4):578-582. DOI: 10.11113/jt.v78.9570

[80] Aina OM, Adetogun AC, Iyiola KA. Heat energy from value-added sawdust briquettes of Albizia Zygia. Ethiopian Journal of Environmental Studies and Management. 2009;**2**(1):42-29. DOI: 10.4314/ejesm.v2i1.43501

[81] Garba B. Overview of Biomass Energy Resources, Technologies and Programmes in Nigeria. Sokoto, Nigeria: SERC-UDUS; 2011. Available from: https://docplayer.net/81086935-Overview-of-biomass-energy-resources-technologies-and-programmes-in-nigeria-serc-udus-sokoto-nigeria.html

[82] Olorunnisola AO. Briquetting of rattan furniture waste. Journal of Bamboo and Rattans. 2004;**3**(2): 133-149; pp. 245-262

[83] Oladeji JT, Enweremadu CC, Olafimihan EO. Conversion of agricultural wastes into biomass briquettes. IJAAAR. 2009;**5**(2):116-123

[84] Oladeji JT, Lucas EB. Densification and fuel characteristics of briquettes produced from corncob. Rep Opinion. 2014;**6**(7):71-76. DOI: marsroj060714

[85] Sambo AS. Renewable energy master plan for Nigeria: An assessment. In: Energy Commission of Nigeria. URL: NIPPS Kuru; 2007

[86] Tripathi AK, Iyer PVR, Kandpal TC. A techno-economic evaluation of biomass briquetting in India. Biomass and Bioenergy. 1998;**14**:479-488. DOI: 10.1016/S0961-9534(97)10023-X

[87] Stolarski MJ, Szczukowski S, Tworkowski J, Krzyzaniak M, Gulczyński P, Mleczek M. Comparison of quality and production cost of briquettes made from agricultural and forest origin biomass. Renewable Energy. 2013;**57**:20-26

[88] Bhattacharya SC, Albina DO, Khaing AM. Effects of selected parameters on performance and emission of biomass-fired cookstoves. Biomass and Bioenergy. 2002;**23**:387-395

[89] Qian K, Kumar A, Patil K, Bellmer D, Wang D, Yuan W, et al. Effects of biomass biomass and gasification conditions on the physiochemical properties of char. Energies. 2013;**6**:3972-3986

[90] Pham KT, Cuong ND, Leon MA. Activities and Achievements of a Biomass Briquetting Project in Vietnam. Vietnam, Hanoi: Institute of Energy IE; 2006. p. 2006

[91] Spalding MA, Sun X, Womer TW. Cost Analysis for Installing New and Optimized Screws for Single-Screw Extrusion Lines. Conference: SPE ANTEC Tech. Papers At: Orlando, FL, 2018

[92] Michels R, Wortberg J. Innovative developments in single-screw extrusion—Possibilities and limitations of high-output extrusion. In: Proceedings of the VDI Conference on "the Single Screw Extruder—Basics and System Optimization". VDI-Verlag Düsseldorf; 1997. pp. 1-21

[93] Bello RS, Onilude MA. Physico-mechanical characteristics of high-density briquettes produced from composite sawdust admixture. Journal of Applied Sciences and Environmental Management. 2020;**24**(5):779-787. DOI: 10.4314/jasem.v24i5.8

[94] Davies RM, Davies OA. Physical and combustion characteristics of briquettes made from water hyacinth and phytoplankton scum as binder. Journal of Combustion. 2013;**2013**(1):549894

[95] Karunanithy C, Wang Y, Muthukumarappan K, Pugalendhi S. Physiochemical chatacterisation of briquettes made from different feedstocks. Biotechnology Research International. 2012;**2012**(1):12. Available from: https://downloads.hindawi.com/archive/2012/165202.pdf

[96] Proskurina S, Heinimö J, Schipfer F, Vakkilainen E. Biomass for industrial applications: The role of torrefaction. Renewable Energy. 2017;**111**:265-274

[97] Onchieku JM. Cost benefit analysis of making charcoal briquettes using screw press machine locally designed and fabricated. International Advanced Research Journal in Science Engineering and Technology. 2018;**5**(6):57-65. DOI: 10.17148/IARJSET.2018.569 57

[98] Sengar SH, Patil SSA, Chendake D. Economic feasibility of briquetted fuel. Global Journal of Researches in Engineering Chemical Engineering. 2013;**13**(1):21-26

[99] Ackerman P, Belbo H, Eliasson L, de Jong A, Lazdins A, Lyons J. The COST model for calculation of forest operations costs. International Journal of Engineering. 2014;**25**:75-81

[100] Spinelli R, Visser RJM. Analyzing and estimating delays in harvester operations. International Journal of Forest Engineering. 2008;**19**:36-41

[101] Laina RR, Esteban ET, Rodríguez SJH. Economic and life cycle assessment of integrated wood and chips harvesting from hybrid poplar plantations in the Genil Valley (Spain). In: Comparison with Chips Harvesting from Poplar SRCs: in Proceedings of the 49th FORMEC Symposium September 4-7, Warsaw, Poland. 2016

[102] Ngubane NF, Oyekola OO. Optimisation of the production of pyrolysed corn stover briquettes and its techno-economic analysis. Waste and Biomass Valorization. 2023;**14**:1333-1354

[103] Chen Y. Biomass to fuels: Thermo-chemical or bio-chemical conversion? Fermentation Technology. 2012;**01**(02):7972

Chapter 7

Succinic Acid: A Sustainable Platform

Lethiwe D. Mthembu and Nirmala Deenadayalu

Abstract

Succinic acid, a crucial C4-dicarboxylic acid, is essential in biological metabolism and serves as a precursor for key industrial chemicals like 1,4-butanediol and tetrahydrofuran. Traditionally derived from fossil fuels, its production raises significant environmental concerns. This chapter explores succinic acid's importance and wide-ranging applications, emphasizing sustainable production methods. Focus is given to microbial fermentation using renewable resources and carbon dioxide. The discussion includes advances in genetic engineering, feedstock selection, and process optimization to improve efficiency. Highlighting the development of bio-based polymers and sustainable industrial practices, the chapter illustrates succinic acid's pivotal role in fostering a greener future for the chemical industry.

Keywords: succinic acid, platform chemical, microbial fermentation, renewable resources, sustainable chemistry

1. Introduction

Succinic acid, a versatile and essential member of the C4-dicarboxylic acid family, is also known as butanedioic acid or amber acid. It is a naturally occurring compound found in humans, animals, plants, and microorganisms, playing a critical role in biological metabolism. As an intermediate in the citric acid cycle, succinic acid is fundamental to energy production and various metabolic processes [1, 2].

Beyond its biological significance, succinic acid is a valuable platform chemical, serving as a precursor for the synthesis of numerous important industrial commodities. These include 1,4-butanediol, tetrahydrofuran, and γ-butyrolactone, which are essential in the production of plastics, resins, solvents, and other materials [3]. **Figure 1** illustrates the diverse applications of succinic acid in various industries.

Currently, the commercial production of succinic acid predominantly relies on fossil fuels, a process that not only depletes finite resources but also releases significant pollutants. Considering increasing environmental concerns and the need for sustainable practices, research has shifted toward producing succinic acid via fermentation from renewable biological resources and carbon dioxide, a major greenhouse gas [2]. This biotechnological approach promises a more environmentally friendly and sustainable method of production, aligning with global efforts to reduce carbon footprints and mitigate climate change.

IntechOpen

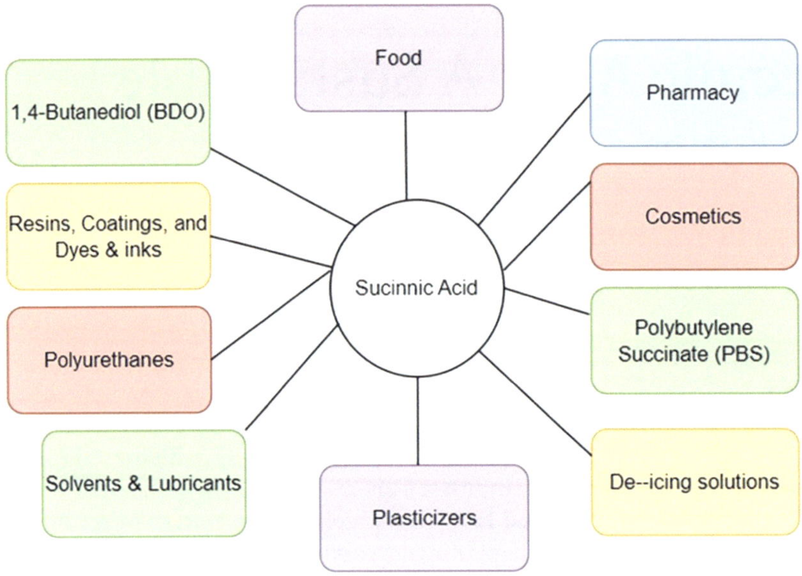

Figure 1.
Illustrates the succinic acid in varies fields.

This chapter aims to provide a comprehensive overview of succinic acid, emphasizing its significance in biological and industrial contexts, current production methods, and the potential of sustainable biotechnological approaches. Through a detailed exploration of microbial fermentation, genetic engineering advancements, and feedstock optimization, we will highlight the prospects of succinic acid as a key player in the transition toward greener and more sustainable chemical production.

2. Biological role and metabolism

2.1 Overview of succinic acid in biological systems

Succinic acid, also known as butanedioic acid or amber acid, is a naturally occurring dicarboxylic acid that plays a fundamental role in the metabolism of living organisms. It is an intermediate in the citric acid cycle (Krebs cycle), which is a crucial metabolic pathway that generates energy through the oxidation of acetate derived from carbohydrates, fats, and proteins. This pathway is essential for the production of adenosine triphosphate (ATP), the primary energy carrier in cells [1].

2.2 Role in the citric acid cycle and overall metabolism

In the citric acid cycle, succinic acid is formed from succinyl-CoA through a reaction catalyzed by the enzyme succinyl-CoA synthetase. Succinic acid is then converted to fumaric acid by the enzyme succinate dehydrogenase. This conversion is coupled with the reduction of ubiquinone to ubiquinol in the mitochondrial membrane, playing a key role in the electron transport chain and oxidative phosphorylation, processes that produce ATP [2].

The citric acid cycle, and specifically the role of succinic acid, is central to the cellular respiration process, enabling organisms to efficiently convert biochemical energy from nutrients into ATP. This energy is vital for various cellular functions, including muscle contraction, nerve impulse propagation, and biosynthesis of essential molecules.

2.3 Succinic acid in human metabolism

In humans, succinic acid levels can influence various physiological processes. For instance, it acts as a signaling molecule in the regulation of glucose homeostasis and insulin secretion. Elevated succinic acid levels have been associated with metabolic disorders such as diabetes and obesity. Moreover, succinic acid and its derivatives have potential therapeutic applications due to their roles in modulating inflammation, oxidative stress, and mitochondrial function [4].

2.4 Succinic acid in plants and microorganisms

In plants, succinic acid is involved in the metabolic pathways that support growth and stress responses. It acts as an intermediate in the glyoxylate cycle, which enables plants to convert fatty acids into carbohydrates during seed germination. In microorganisms, succinic acid serves as a key metabolite in anaerobic fermentation processes. Certain bacteria and fungi can produce succinic acid from various carbon sources, making them valuable for industrial biotechnological applications [5].

2.5 Ecological and environmental significance

Succinic acid also plays an ecological role in the environment. Soil microorganisms utilize succinic acid in the decomposition of organic matter, contributing to nutrient cycling and soil health. Additionally, succinic acid can act as a carbon source for certain bacteria involved in bioremediation processes, aiding in the cleanup of contaminated environments [6].

3. Industrial applications

3.1 Detailed analysis of succinic acid as a precursor for various chemicals

Succinic acid is a highly versatile platform chemical that serves as a precursor for the synthesis of numerous valuable industrial products. Its molecular structure, characterized by two carboxyl groups, enables a variety of chemical transformations, making it a crucial intermediate in the production of many compounds used in diverse industries.

3.1.1 1,4-Butanediol (BDO)

1,4-Butanediol (BDO) is a primary chemical produced from succinic acid and is utilized in the manufacture of plastics, elastic fibers, and polyurethanes. The hydrogenation of succinic acid yields BDO through a catalytic process that involves the reduction of the carboxyl groups to hydroxyl groups. BDO is a vital component in producing thermoplastic polyurethanes, engineering plastics like polybutylene terephthalate (PBT), and elastic fibers such as Spandex [5].

3.1.2 Tetrahydrofuran (THF)

Tetrahydrofuran (THF) is another significant chemical derived from succinic acid via the intermediate BDO. THF is produced through the dehydration of BDO and is extensively used as an industrial solvent, particularly in the production of polyvinyl chloride (PVC) and as a precursor to various polymers. THF is known for its ability to dissolve a wide range of substances, making it invaluable in pharmaceutical and chemical synthesis [4].

3.1.3 Gamma-butyrolactone (GBL)

Gamma-Butyrolactone (GBL) is synthesized from succinic acid through a series of catalytic reactions. GBL serves as a solvent, reagent, and intermediate in the production of chemicals such as pyrrolidones and herbicides. It is also used in the synthesis of pharmaceuticals, including certain anesthetics and tranquilizers, due to its chemical stability and reactivity [7].

3.1.4 Bio-based polymers

Succinic acid is a key monomer in the production of bio-based polymers, particularly polybutylene succinate (PBS). PBS is a biodegradable polyester with properties similar to polyethylene and polypropylene, making it suitable for applications in packaging, agricultural films, and disposable products. The polymerization of succinic acid with 1,4-butanediol results in PBS, highlighting the importance of succinic acid in developing sustainable and environmentally friendly materials [6].

3.1.5 Succinic anhydride

Succinic anhydride is produced through the dehydration of succinic acid and is used as a chemical intermediate in the production of alkyd resins, plasticizers, and surfactants. It also finds applications in the manufacture of pharmaceuticals, agrochemicals, and as a curing agent for epoxy resins. Succinic anhydride's reactivity with alcohols and amines makes it valuable in creating various functionalized compounds [1].

3.1.6 Pharmaceuticals and nutraceuticals

Succinic acid and its derivatives are used in the pharmaceutical and nutraceutical industries due to their biocompatibility and therapeutic properties. Succinic acid acts as a key ingredient in formulations aimed at treating metabolic disorders, enhancing physical performance, and promoting overall health. Its role in modulating mitochondrial function and reducing oxidative stress underlines its potential in developing novel therapeutic agents [2].

4. Current production methods

4.1 Fossil fuel-based production

4.1.1 Traditional methods using fossil fuels as raw materials

4.1.1.1 Paraffin oxidation

This method uses a calcium or manganese catalyst to oxidize paraffin, producing a mixture of dicarboxylic acids, including succinic acid. The process, industrialized since the 1930s, offers low yield and purity of succinic acid, necessitating extensive purification steps [8, 9].

4.1.1.2 Catalytic hydrogenation

In this method, maleic acid or maleic anhydride is hydrogenated to produce succinic acid with higher yield and purity. This process can be performed homogeneously or heterogeneously depending on the catalyst. Despite its efficiency, the process is expensive, complex, and environmentally taxing due to the reliance on fossil fuels and high energy requirements [10].

4.1.2 Environmental impact and sustainability issues

Traditional fossil fuel-based methods for succinic acid production have significant environmental drawbacks. Paraffin oxidation, while capable of producing succinic acid, contributes to greenhouse gas emissions and relies on nonrenewable resources, resulting in low sustainability. The catalytic hydrogenation method, although more efficient, also suffers from environmental concerns due to the energy-intensive nature of the process and the use of fossil-derived materials [10]. These methods are increasingly viewed as unsustainable, especially in the context of rising crude oil prices and stringent environmental regulations [9].

4.2 Biotechnological approaches

4.2.1 Fermentation processes using renewable biological resources

Metabolic pathways: The production of succinic acid via fermentation involves metabolic pathways in microorganisms such as the reductive branch of the TCA cycle, which is active under anaerobic conditions. Microorganisms like *Actinobacillus succinogenes* and genetically engineered *Escherichia coli* are commonly employed. These processes utilize renewable biomass feedstocks, such as corn stover, cane molasses, and glycerol, offering a sustainable alternative to fossil fuel-based methods [9, 11].

Microbial fermentation: Fermentation processes use renewable resources like glucose, glycerol, and agricultural by-products. For example, *Actinobacillus succinogenes* can efficiently convert corn stover hydrolyzates into succinic acid. Other feedstocks like whey, glycerol, and lignocellulosic biomass are also explored, providing a sustainable alternative to fossil fuels [11].

4.2.2 Use of carbon dioxide, a greenhouse gas, in the production of succinic acid

Utilization of CO_2: An environmental advantage of biotechnological succinic acid production is the use of CO_2 as a feedstock. In the fermentation process, CO_2 is assimilated and contributes to succinic acid formation, potentially reducing overall greenhouse gas emissions. This method not only helps mitigate CO_2 emissions but also adds value by converting CO_2 into a valuable chemical product. This biotechnological approach exemplifies how advances in this field can lead to more sustainable industrial processes [9, 11].

5. Sustainable production techniques

5.1 Microbial fermentation

Microbial fermentation has emerged as a sustainable and efficient method for producing biochemicals, including succinic acid. Through the metabolic pathways of genetically engineered microorganisms, renewable feedstocks are converted into valuable bioproducts, offering an environmentally friendly alternative to petrochemical methods [9].

5.1.1 Overview of microorganisms used for succinic acid production

Several microorganisms have been employed in the industrial production of succinic acid due to their ability to efficiently utilize diverse carbon sources. Key organisms include:

5.1.1.1 Actinobacillus succinogenes

A well-studied anaerobic bacterium for succinic acid production, capable of utilizing various carbon sources like agro-industrial waste. It is known for its robustness and high yield in industrial fermentation processes [12].

5.1.1.2 Escherichia coli

Engineered strains of *E. coli* are frequently used due to their flexible metabolic pathways and ability to ferment glucose into succinic acid. Several strategies have been developed to block by-products and improve yields [13, 14].

5.1.1.3 Mannheimia succiniciproducens

This bacterium has been optimized for homo-succinic acid production using metabolic engineering techniques that direct more carbon flux toward succinic acid and limit the formation of by-products [15].

5.1.1.4 Basfia succiniciproducens

Another rumen bacterium capable of producing succinic acid efficiently from glucose and sucrose through optimized metabolic pathways [16].

5.1.2 Advances in genetic engineering to enhance production efficiency

Recent advancements in genetic engineering have further improved microbial succinic acid production efficiency. The key areas of development include:

- *Metabolic pathway optimization*: Genetic modifications to suppress by-product formation pathways (e.g., acetate and lactate) and redirect carbon flux to succinic acid production have significantly improved yields [17]

- *Gene overexpression*: Enhancing the expression of genes like phosphoenolpyruvate carboxylase (PEPC) has increased succinic acid yields by optimizing key intermediates in the TCA cycle [18].

- *CRISPR-Cas9 technology*: The use of CRISPR-Cas9 has enabled rapid and precise genetic editing, allowing the simultaneous modification of multiple genes involved in succinic acid metabolism [19].

- *Cofactor regeneration engineering*: Efficient regeneration of cofactors such as NADH/NAD+ has been engineered into strains to improve the reductive pathway, further increasing fermentation efficiency and succinic acid yields [15].

- *Synthetic biology tools*: Modular genetic circuits and synthetic pathways have been constructed using synthetic biology techniques to control metabolic flux and enhance the robustness of production strains under industrial conditions [17].

These advancements are making microbial fermentation a highly efficient and sustainable method for succinic acid production, reducing reliance on petrochemical-based methods.

5.2 Feedstock selection

5.2.1 Use of Lignocellulosic biomass and other renewable feedstocks

Lignocellulosic biomass, including sugarcane bagasse, has gained attention as a viable and sustainable feedstock for succinic acid production due to its abundance and renewability. Several studies have explored the potential of these feedstocks to replace fossil-based sources, and significant advancements have been made in the bioconversion of lignocellulose.

Sugarcane bagasse is one of the most utilized lignocellulosic biomasses for succinic acid production. Studies have shown that the use of ultrasonically pretreated sugarcane bagasse with dilute acid hydrolysis can yield a total sugar concentration of 43.9 g/L, leading to 23.7 g/L of succinic acid production [20]. Another process using sugarcane bagasse for co-production of succinic acid and ethanol, utilizing both glucose and xylose from hydrolysates, achieved concentrations of 22.1 g/L for succinic acid and 22.0 g/L for ethanol [21].

The co-fermentation of sugarcane bagasse can enhance the economic feasibility of using lignocellulosic biomass. Co-producing multiple products, such as bioethanol and succinic acid, has shown promise in increasing the yield and sustainability of biomass biorefining processes [22].

Other biomasses such as corn stover, switchgrass, and Arundo donax have also been tested for succinic acid production. *Basfia succiniciproducens* was used to produce succinic acid from hydrolysates of Arundo donax, yielding 9.4 g/L of succinic acid [23]. Another study on corn stover demonstrated succinic acid titers reaching 30 g/L using microbial fermentation [24].

5.2.2 Economic and environmental benefits of using renewable resources

The use of renewable feedstocks such as sugarcane bagasse and other lignocellulosic biomasses offers significant economic and environmental benefits.

5.2.2.1 Economic viability

Sugarcane bagasse, as a low-cost and abundant waste product from the sugarcane industry, can substantially reduce production costs when used as a feedstock for succinic acid production. By co-producing other value-added chemicals such as polyhydroxybutyrate (PHB) and electricity, the overall profitability of biorefineries can be enhanced. For example, a study found that the co-production of succinic acid and PHB in a sugarcane bagasse biorefinery resulted in an internal rate of return (IRR) of 24.1% [25].

5.2.2.2 Environmental sustainability

Utilizing renewable feedstocks like lignocellulose contributes to lower greenhouse gas emissions compared to fossil-based processes. A study demonstrated that using Arundo donax for succinic acid production reduced net global warming potential by 55% compared to fossil adipic acid production [26]. Additionally, the recycling of CO_2 during the fermentation process, such as in the co-fermentation of ethanol and succinic acid from sugarcane bagasse, further reduces CO_2 emissions [21].

5.3 Process optimization

5.3.1 Technological advancements in fermentation processes

Significant advancements have been made in optimizing fermentation processes to increase the efficiency and yield of succinic acid production. Some of these advancements include:

5.3.1.1 Continuous fermentation systems

Continuous fermentation using engineered strains like Mannheimia succiniciproducens LPK7 has shown promising results. This system achieved a succinic acid yield of 1.77 g/L/h with reduced by-product formation. Continuous processes allow for higher productivity compared to batch fermentation, making them more suitable for industrial-scale production [27].

5.3.1.2 Cell-recycled continuous fermentation (CRCF)

CRCF has been developed to enhance succinic acid productivity by maintaining high cell densities, overcoming the limitations of cell washout during continuous fermentation. This approach has resulted in a significant increase in volumetric productivity to 3.86 g/L/h, which is 5.1 times higher than traditional fermentation methods [28].

5.3.1.3 Use of metabolically engineered strains

Advances in metabolic engineering have led to the development of strains that exhibit higher succinic acid yields. For example, Escherichia coli strains engineered to overexpress formate dehydrogenase have increased succinic acid production by enhancing NADH availability, crucial for the reductive carboxylation of pyruvate to succinic acid [29].

5.3.2 Strategies for improving yield and reducing production costs

Several strategies have been employed to improve succinic acid yields and reduce production costs:

5.3.2.1 Optimizing Feedstocks

Utilizing low-cost renewable feedstocks, such as sugarcane molasses, has been shown to reduce production costs significantly. In anaerobic fermentation using *Actinobacillus succinogenes*, succinic acid concentrations of 55.2 g/L were achieved using fed-batch fermentation, which also minimized substrate inhibition effects [30].

5.3.2.2 Electro-membrane extraction

Recovery processes are critical in reducing overall production costs. Membrane electrolysis and electro-membrane extraction have shown promise in efficiently separating succinic acid from fermentation broths. These processes enhance energy efficiency and reduce the cost of downstream purification [31].

5.3.2.3 Immobilized cell fermentation

The use of immobilized cells can lower costs by enabling multiple fermentation cycles. This strategy has been successfully applied in glycerol-based succinic acid production, achieving high succinic acid yields while allowing for cell reuse across fermentation batches [32].

5.3.2.4 pH regulation and neutralizers

Optimizing pH during fermentation can reduce costs by lowering the need for expensive neutralizers like NaOH. For instance, a two-stage pH regulation strategy in Yarrowia lipolytica fermentation reduced NaOH consumption by 43%, improving the overall process economy [33].

6. Applications and future prospects

6.1 Bio-based polymers and materials

6.1.1 Use of succinic acid in the production of bio-based polymers

Succinic acid has become a key platform chemical in the production of bio-based polymers due to its versatility and the growing demand for sustainable alternatives to petroleum-based plastics. The application of succinic acid spans the development of various polymers, including poly(butylene succinate) (PBS), polyesters, and polyamides.

6.1.1.1 Poly(butylene succinate) (PBS)

PBS is one of the most important bioplastics derived from succinic acid and 1,4-butanediol. PBS is biodegradable and possesses properties similar to conventional plastics, making it suitable for a wide range of applications, including packaging, agriculture, and biodegradable consumer goods [34, 35].

6.1.1.2 Polyesters and polyamides

Succinic acid is also used in the synthesis of polyesters and polyamides, which are biodegradable and used in a variety of industries, including textiles, automotive parts, and electronics. These materials can replace petrochemical-derived polymers and reduce the environmental footprint [36].

6.1.1.3 Bio-based plasticizers

Succinic acid derivatives, such as bis(2-ethylhexyl) succinate, have been explored as bio-based plasticizers in polymers like poly(vinyl chloride) (PVC), showing promise as sustainable alternatives to petrochemical plasticizers [37].

6.1.2 Potential applications in sustainable materials and bioplastics

The use of succinic acid in the production of bio-based polymers has opened up numerous applications in sustainable materials, particularly in the area of biodegradable plastics. These materials are gaining popularity due to their reduced environmental impact and ability to biodegrade in natural environments.

6.1.2.1 Biodegradable plastics

PBS, produced from succinic acid, is widely used in the manufacture of biodegradable plastics for single-use products such as food packaging, disposable cutlery, and agricultural films. The material's ability to decompose in soil and compost environments makes it a promising alternative to conventional plastics [38].

6.1.2.2 Sustainable coatings

Succinic acid is also used in the production of bio-based coatings, especially for environmentally friendly applications such as food-safe packaging. Coatings developed from succinic acid-based polymers exhibit excellent mechanical properties and chemical resistance [39].

6.1.2.3 Advanced biocomposites

The combination of succinic acid-derived polymers with natural fibers has led to the development of biocomposites for automotive, construction, and electronics industries. These materials are lighter, biodegradable, and offer a sustainable alternative to traditional composites [40].

6.2 Chemical industry integration

6.2.1 Integration of bio-based succinic acid production into existing chemical manufacturing processes

The integration of bio-based succinic acid production into existing chemical manufacturing processes is increasingly viable due to advancements in biotechnology and bio-refinery systems. Succinic acid, as a platform chemical, is compatible with various chemical transformation processes traditionally dependent on petroleum-based feedstocks, making it attractive for the chemical industry.

6.2.1.1 Energy and greenhouse gas savings

Studies show that bio-based succinic acid production using methods like low pH yeast fermentation has a significantly lower greenhouse gas (GHG) emission profile compared to traditional petrochemical-based methods. Integrating bio-based succinic acid into existing chemical processes offers both energy savings and environmental benefits, particularly in reducing nonrenewable energy use (NREU) [41].

6.2.1.2 Bioreactor and process integration

The development of advanced bioreactors, including those that incorporate continuous and integrated fermentation and separation systems, has been pivotal in optimizing succinic acid production for chemical industry integration. These technological advancements are designed to reduce costs, streamline processes, and improve yield, making bio-succinic acid a more economically feasible alternative [42].

6.2.1.3 Economic competitiveness

The combination of metabolic engineering with advanced downstream processing, such as reactive extraction and electrodialysis, makes bio-based succinic acid more competitive with its petrochemical counterparts. These approaches enhance process efficiency, lower production costs, and reduce environmental impacts, supporting large-scale industrial integration [43].

6.2.2 Future trends and potential for large-scale industrial adoption

The future of bio-based succinic acid production looks promising, with trends indicating a shift toward larger-scale industrial adoption. Several factors are contributing to this movement:

6.2.2.1 Process intensification and scale-up

As bio-based succinic acid production technology matures, scale-up efforts are accelerating. Companies are moving from pilot to full-scale production, making bio-succinic acid more commercially viable. Continuous improvements in fermentation processes and metabolic engineering of microorganisms like *E. coli* and Mannheimia succiniciproducens are key to achieving higher yields and reduced costs [44].

6.2.2.2 Market demand for biodegradable polymers

The demand for biodegradable plastics, such as poly(butylene succinate) (PBS) produced from bio-based succinic acid, is growing. This trend is driven by the increasing push for sustainability and the reduction of plastic pollution. Bio-succinic acid is becoming a critical feedstock in the production of bioplastics, which are gaining acceptance in industries such as packaging, automotive, and agriculture [38].

6.2.2.3 Sustainability and circular economy

As the world moves toward a circular economy, bio-based succinic acid plays a key role in reducing reliance on fossil fuels and minimizing environmental impacts. The ability of bio-based succinic acid production to integrate with existing chemical production systems, coupled with its application in a wide range of industries, makes it a strong candidate for large-scale adoption [45].

7. Conclusions

This chapter has comprehensively explored the pivotal role of succinic acid as a sustainable platform chemical, focusing on its biological significance, industrial applications, and the advancements in sustainable production methods. Succinic acid, as a key intermediary in biological metabolism and an essential precursor for numerous industrial chemicals, stands at the intersection of environmental sustainability and industrial innovation. The transition from traditional fossil fuel-based methods to biotechnological approaches for succinic acid production marks a significant shift toward greener, more sustainable industrial practices.

The utilization of microbial fermentation, genetic engineering, and renewable feedstocks such as lignocellulosic biomass demonstrates the potential of bio-based succinic acid to reduce environmental impact while maintaining industrial efficiency. Microbial fermentation, particularly using engineered strains, offers a viable and scalable method for producing high yields of succinic acid, with significant advances in optimizing metabolic pathways, enhancing cofactor regeneration, and leveraging CRISPR technology. These advancements are crucial for the development of biorefineries that produce succinic acid at commercial scales, potentially replacing petrochemical methods.

The integration of bio-based succinic acid into the broader chemical industry is progressing rapidly, with significant reductions in greenhouse gas emissions and energy use. As the demand for biodegradable polymers such as poly(butylene succinate) increases, succinic acid is positioned as a key feedstock in the production of sustainable plastics, biocomposites, and bio-based materials. These developments align with global trends toward a circular economy and the reduction of plastic waste, further highlighting the critical role of succinic acid in fostering sustainable industrial processes.

Succinic acid's dual role as a biological metabolite and a versatile platform chemical place it at the forefront of efforts to transition to a more sustainable, bio-based chemical industry. With ongoing technological innovations in microbial fermentation, genetic engineering, and feedstock utilization, the large-scale adoption of bio-based succinic acid production is not only feasible but also essential for a greener industrial future. The chemical industry's integration of succinic acid into existing production processes presents a unique opportunity to enhance environmental sustainability without sacrificing economic viability.

Acknowledgements

The authors are grateful to (i) National Research Foundation (NRF) UID 138554 for financial support. (ii) Durban University of Technology (GOOT) for financial support and facilities to conduct this research.

Conflict of interest

The authors declare no conflict of interest.

Author details

Lethiwe D. Mthembu* and Nirmala Deenadayalu
Department of Chemistry, Durban University of Technology, Durban, South Africa

*Address all correspondence to: lethiwem@dut.ac.za

References

[1] Zeikus JG, Jain MK, Elankovan P. Biotechnology of succinic acid production and markets for derived industrial products. Applied Microbiology and Biotechnology. 1999;**51**(5):545-552

[2] Jiang M, Zhang H, Lan H, Lin S. Fermentative production of succinic acid: An emerging technology to replace the traditional petrochemical processes. BioMed Research International. 2013;**2013**:723412

[3] McKinlay JB, Vieille C, Zeikus JG. Prospects for a bio-based succinate industry. Applied Microbiology and Biotechnology. 2007;**76**(4):727-740

[4] Cukalovic A, Stevens CV. Feasibility of production methods for succinic acid derivatives: A marriage of renewable resources and chemical technology. Biofuels, Bioproducts and Biorefining. 2008;**2**(6):505-529

[5] Song H, Lee SY. Production of succinic acid by bacterial fermentation. Enzyme and Microbial Technology. 2006;**39**(3):352-361

[6] Vardon DR, Franden MA, Johnson CW, Beckham GT. Chemical and biological catalysis for plastics upcycling. Nature Catalysis. 2015;**1**(12):930-943

[7] Bechthold I, Bretz K, Kabasci S, Kopitzky R, Springer A. Succinic acid: A new platform chemical for biobased polymers from renewable resources. Chemical Engineering and Technology. 2008;**31**(5):647-654

[8] Beauprez JJ, De Mery M, Soetaert WK. Microbial succinic acid production: Natural versus metabolic engineered producers. Process Biochemistry. 2010;**45**(7):1103-1114

[9] Cheng KK, Zhao XB, Zeng J, Zhang JA. Biotechnological production of succinic acid: Current state and perspectives. Biofuels, Bioproducts and Biorefining. 2012;**6**(3):302-318

[10] Singh A, Soh K, Hatzimanikatis V, Gill R. Manipulating redox and ATP balancing for improved production of succinate in *E. Coli*. Metabolic Engineering. 2010;**13**:76-81

[11] Jiang M, Zhang H, Lan S, Luo Y. The utilization of carbon dioxide as a feedstock in the production of succinic acid. Biotechnology Advances. 2013;**31**(3):615-620

[12] Putri DN, Sahlan M, Montastruc L, Meyer M, Négny S, Hermansyah H. Progress of Fermentation Methods for Bio-Succinic Acid Production Using Agro-Industrial Waste by *Actinobacillus succinogenes*. Energy Reports. Amsterdam, Netherlands: Elsevier; 2020

[13] Zhao J, Hua B, Wang Y, Liu Z, Wang J, Zhou S. Fermentation characteristics of engineered *Escherichia coli* for succinic acid production. In: 2013 International Conference on Materials for Renewable Energy and Environment. Vol. 3. Piscataway, NJ, USA: IEEE; 2013. pp. 982-985

[14] Zhang Y, Wang J, Wang D, Ji F. Advancement in genetic engineering for production of succinic acid by *Escherichia coli*. China Biotechnology. 2009;**29**(07):108-117

[15] Lee JW, Yi J, Kim T, Choi S, Ahn J, Song H, et al. Homo-succinic acid production by metabolically engineered Mannheimia succiniciproducens. Metabolic Engineering. 2016;**38**:409-417

[16] Lange A, Becker J, Schulze D, Cahoreau E, Portais J, Haefner S, et al. Bio-based succinate from sucrose: High-resolution 13C metabolic flux analysis and metabolic engineering of the rumen bacterium Basfia succiniciproducens. Metabolic Engineering. 2017;**44**:198-212

[17] Kim W, Ahn J, Kim H, Kim T, Lee S. Metabolic engineering of Mannheimia succiniciproducens for succinic acid production based on elementary mode analysis with clustering. Biotechnology Journal. 2017;**12**(2):1-21

[18] Ahn J, Jang Y, Lee S. Production of succinic acid by metabolically engineered microorganisms. Current Opinion in Biotechnology. 2016;**42**:54-66

[19] Guarnieri M, Chou Y, Salvachúa D, Mohagheghi A, John PC, Peterson D, et al. Metabolic engineering of *Actinobacillus succinogenes* provides insights into succinic acid biosynthesis. Applied and Environmental Microbiology. 2017

[20] Xi Y, Dai W, Xu R, Zhang J, Chen K, Jiang M, et al. Ultrasonic pretreatment and acid hydrolysis of sugarcane bagasse for succinic acid production using *Actinobacillus succinogenes*. Bioprocess and Biosystems Engineering. 2013;**36**:1779-1785

[21] Xu C, Alam M, Wang Z, Peng Y, Xie C, Gong W, et al. Co-fermentation of succinic acid and ethanol from sugarcane bagasse based on full hexose and pentose utilization and carbon dioxide reduction. Bioresource Technology. 2021;**339**:125578

[22] Bu S, Li Q, Gao X, Xin F, Ma C, Jiang M. Co-production of bioethanol and succinic acid from sugarcane bagasse by engineered *Escherichia coli* under anaerobic conditions. Bioresource Technology. 2019;**273**:370-378

[23] Ventorino V, Robertiello A, Cimini D, Pepe O, Schiraldi C. Basfia succiniciproducens for succinic acid production from Arundo donax hydrolysates: A biorefinery approach. Bioresource Technology. 2017;**243**:940-947

[24] Salvachúa D, Smith H, John PC, Mohagheghi A, Peterson D, Black B, et al. Succinic acid production from lignocellulosic hydrolysate by Basfia succiniciproducens. Bioresource Technology. 2016;**214**:558-566

[25] Nieder-Heitmann M, Haigh K, Görgens J. Process design and economic evaluation of integrated, multi-product biorefineries for the co-production of bio-energy, succinic acid, and polyhydroxybutyrate (PHB) from sugarcane bagasse and trash lignocelluloses. Biofuels. 2019;**10**(2):265-278

[26] Zucaro A, Fiorentino G, Sepe I, Ulgiati S. Life cycle assessment of succinic acid production from Arundo donax via a biorefinery process. Journal of Cleaner Production. 2017;**142**:3482-3490

[27] Oh IJ, Lee H-W, Park C, Lee S, Lee J. Succinic acid production from continuous fermentation process using Mannheimia succiniciproducens LPK7. Journal of Microbiology and Biotechnology. 2008;**18**(5):908-912

[28] Kim SY, Park SO, Yeon JY, Chun G. Development of a cell-recycled continuous fermentation process for enhanced production of succinic acid by high-yielding mutants of *Actinobacillus succinogenes*. Biotechnology and Bioprocess Engineering. 2020;**26**:125-136

[29] Balzer GJ, Thakker C, Bennett G, San K. Metabolic engineering of *Escherichia coli* to minimize byproduct

formate and improve succinate productivity through increasing NADH availability by heterologous expression of NAD(+)-dependent formate dehydrogenase. Metabolic Engineering. 2013;**20**:1-8

[30] Liu Y, Zheng P, Sun Z, Ni Y, Dong J, Zhu L. Economical succinic acid production from cane molasses by *Actinobacillus succinogenes*. Bioresource Technology. 2008;**99**(6):1736-1742

[31] Malanca AA, Mancini E, Yusuf M, Khensir G, Mansouri S, Skiadas I, et al. Variables and mechanisms affecting electro-membrane extraction of bio-succinic acid from fermentation broth. Membranes. 2022;**12**(2):1-15

[32] Aziati AHN, Sakinah A. Efficient production of succinic acid in immobilized fermentation with crude glycerol from *Escherichia coli*. Journal of Food Science. 2017;**2**(1):110-118

[33] Stylianou E, Pateraki C, Ladakis D, Damala C, Vlysidis A, Latorre-Sánchez M, et al. Bioprocess development using organic biowaste and sustainability assessment of succinic acid production with engineered *Yarrowia* lipolytica strain. Biochemical Engineering Journal. 2021;**174**:108099

[34] Xu J, Guo B. Microbial succinic acid, its polymer poly(butylene succinate), and applications. Biotechnology Advances. 2010;**28**(3):347-388

[35] Thakur VK, Thakur MK, Raghavan P, Kessler MR. Biodegradable polymer blends and composites from renewable resources. Journal of Cleaner Production. 2021;**21**:140-154

[36] Bechthold I, Bretz K, Kabasci S, Kopitzky R, Springer A. Succinic acid: A new platform chemical for biobased polymers from renewable resources. Chemical Engineering and Technology. 2008;**31**(6):647-654

[37] Chaudhary BI, Nguyen B-D, Smith PB, Sunday N, Luong M, Zamanskiy A. Bis (2-ethylhexyl) succinate in mixtures with epoxidized soybean oil as bio-based plasticizers for poly(vinyl chloride). Polymer Engineering and Science. 2015;**55**(3):634-640

[38] Patel M, Bechu A, Villegas J, Bergez-Lacoste M, Yeung K, Murphy R. Second-generation bio-based plastics: Non-renewable energy and GHG balance of succinic acid-based plastic end products. Biofuels. 2018;**12**(1):1-14

[39] Noordover B, van Staalduinen VG, Duchateau R, Koning C, van Benthem RV, Mak M, et al. Co- and terpolyesters based on isosorbide and succinic acid for coating applications. Biomacromolecules. 2006;**7**(12):3406-3416

[40] Ioannidou S, Pateraki C, Ladakis D, Papapostolou H, Tsakona M, Vlysidis A, et al. Sustainable production of bio-based chemicals and polymers via integrated biomass refining. Bioresource Technology. 2020;**306**:123093

[41] Cok B, Tsiropoulos I, Roes AL, Patel MK. Succinic acid production derived from carbohydrates: An energy and greenhouse gas assessment of a platform chemical toward a bio-based economy. Biofuels, Bioproducts and Biorefining. 2014;**8**(1):16-29

[42] Ferone M, Raganati F, Olivieri G, Marzocchella A, Salatino P. Bioreactor design and bioprocess integration for biobased succinic acid production. Journal of Industrial Microbiology and Biotechnology. 2019;**46**(9):1231-1241

[43] Morales M, Tretbar S, Dumon I. Economic and environmental

assessment of succinic acid production from lignocellulosic biomass using different downstream separation technologies. Bioresource Technology. 2016;**214**:558-567

[44] Jansen ML, van Gulik WM. Towards large scale fermentative production of succinic acid. Current Opinion in Biotechnology. 2014;**30**:190-197

[45] Bukhari S. The role of bio-based chemicals in a circular economy: Focus on succinic acid production and applications. Journal of Sustainable Chemistry and Circular Economy. 2023;**4**(2):150-165

Chapter 8

Unleashing the Potential of Portuguese Agroforestry Biomass: Extraction, Characterization, and Valorization of Biomolecules

Bruno Medronho, Hugo Duarte, Inês Mansinhos, João Brás, Ana Amorim, Isabela dos Anjos, Maria José Aliaño-González, Raquel Rodríguez-Solana, Luís Alves, Solange Magalhães, Catarina Fernandes, Sandra Gonçalves and Anabela Romano

Abstract

In this chapter, we explore the vast potential of Portuguese agroforestry biomass as a sustainable source of valuable biomolecules. Focusing on the extraction, characterization, and valorization of compounds, such as polysaccharides, phenolics, and essential oils from native species like carob, cork oak, aromatic plants or maritime pine, we highlight their applications in food, pharmaceutical, and innovative bio-based materials. Emphasis is placed on eco-friendly extraction techniques and innovative functionalization methods to enhance the physicochemical properties of different biomolecules, thus expanding their applicability and potential commercial relevance. By highlighting the unexploited diversity of Portuguese agroforestry biomass, we can pave the way for an innovative and, sustainable platform that not only drive economic growth but also preserve biodiversity, reduce waste, and ensure a greener and more prosperous future.

Keywords: Portuguese agroforestry, biomolecule extraction and characterization, bio-based materials, polyphenols, essential oils, locust beam gum, lignocellulose

1. Introduction

Environmental concerns associated with the threat of climate change, environmental degradation, exponential growth of the global population, and scarcity of natural resources have strengthened the emphasis on sustainability. In this context, the circular economy concept that engages the use of renewable bioresources to produce goods and services has gained notoriety in the last years since it aligns with

IntechOpen

sustainability aspirations [1, 2]. Agroforestry biomass, comprising biologically derived products, including waste residues, from forestry and agriculture, appears as an excellent alternative to fossil resources, contributing to minimize the severe impact of non-sustainable fossil-based feedstock [1]. Among the varied potentialities, biomass from forest and agriculture is a key source of valuable biomolecules, such as polysaccharides, with many industrial applications [3]. At the same time, bioeconomy shows promise; it faces many challenges, including the discovery of new feedstocks and optimizing the extraction and characterization of these biomolecules. The exploration of agroforestry biomass can not only fuel the bioeconomy but also foster sustainable farming practices and rural development, establishing a mutually beneficial association.

The Portuguese territory is rich in raw materials from agroforestry resources. According to the last Land Cover and Use Map, about 92% of the territory has agricultural, forestry, and agroforestry occupation; bush and forest represent 51% of the territory, whereas 26% of the territory is used for agriculture. In addition to being a guarantor of biodiversity and a significant carbon sink, forest represents a substantial support for the Portuguese economy [4]. Furthermore, agriculture and associated agro-food industries are important economic sectors in the country. The diverse edaphoclimatic and agroecological conditions in the different regions of Portugal strongly affects plant distribution along the country and allows the cultivation of different agricultural products in each region. The dominant tree species in Portuguese forest are (by this order) maritime pine, eucalyptus, cork-oak, holm-oak, stone pine, oaks, chestnut, acacia, and carob [5]. The Portuguese Flora are also very rich herbaceous plants and shrubs, some of them with aromatic and medicinal characteristics (~500 species) [6].

In addition to the large amounts of residues generated from the forest, agricultural waste has increased daily, mainly because of the increasing demand of the fast-growing population. In fact, the disposal of agricultural waste is currently one of the major environmental issues [6, 7]. Untreated waste contributes to greenhouse gas emissions resulting in undesirable gaseous byproducts that potentially contribute to climate change [7]. Thus, a timely and adequate utilization (and valorization) of this biomass is fundamental for agricultural sustainability and human-food and health security. Agrowastes include nonedible parts of plants, postharvest or postprocessing food residues, such as leaves, peel, seed, or pulp from fruits/legumes, or cereals [6]. These are promising raw materials whose exploration can represent an additional sustainable income to local producers in addition to the environmental benefits inherent to their utilization [8]. Although the main use of forest residues is to produce energy, the Portuguese forest is a reservoir of plant species in which products and byproducts are relevant sources of valuable molecules. The Portuguese Government established strategies to promote the sustainable use and valorization of renewable resources and biomass until 2030 [9]. Agroforestry biomass is mainly composed of organic material and can be divided into lignocellulosic and non-lignocellulosic biomass [10]. Lignocellulosic biomass consists of cellulose, hemicellulose, and lignin, while non-lignocellulosic materials include structural components (sugar, lipids, and proteins) and other organic compounds like secondary metabolites including bioactive compounds. In the last years, consumers have increasingly demanded biodegradable, petroleum-free, and safe products from natural origin. Agroforestry biomass is a key source of biomolecules with many industrial applications that can meet these requirements with improved performance. Furthermore, as we will see through this chapter, biotechnology and nanotechnology, particularly by using eco-friendly and cost-effective methodologies, can play a key role in the bioconversion of agroforestry biomass into added-value unique materials for the pharmaceutical sector, biopesticides, biostimulants and biofuels for the agriculture sector, and nutraceuticals or new food products [11]. Given the great availability of agroforestry biomass in Portugal, its valorization has been gaining a relevant role in the last years. Hence, in this chapter, we provide an overview of the potentialities of Portuguese agroforestry biomass as a primary feedstock for the extraction of relevant biomolecules that could be suitable for the agri-food-pharma sectors. Emphasis is provided to strategic raw materials in the national context, such as carob byproducts, aromatic plants biomass, and lignocellulosics.

2. From carob to pine: Biomolecule diversity in Portugal

Portugal's flora is rich and varied, reflecting the diverse environmental conditions and influences of different biogeographical regions. Along with its high diversity, Portugal's flora continues to expand with recent additions to the country's species list. Notable discoveries include, for instance, *Carduncellus cuatrecasasii* and *Eryngium aquifolium* in the Alentejo [12], *Onobrychis caput-galli* in the Algarve [13], the curious *Lesleya ceriacoi*, a previously unknown gymnosperm species discovered from a 300-million-year-old fossil in the herbarium of the Museum of Natural History and Science of the University of Porto [14]. More recently, species from a family thought to be extinct, *Pseudofrenelopsis zlatkoi* [15] and *Pseudofrenelopsis dinisii* [16], were discovered in the Lusitanian Basin of western Portugal. As the list is constantly expanding, new opportunities for science are emerging, both in terms of the main compounds obtained from plants, as well as secondary compounds and byproducts.

It is known that plants are rich in various bioactive phytochemical compounds, and these compounds can be extracted from various parts such as leaves, bark, seeds, flowers, roots, and pulp. However, as we will explore in the following sections, unleashing the full potential of new biomolecules for various industrial applications relies on effective extraction, thorough analysis, and, in some cases, strategic chemical modifications. Several bioactive compounds present in plants have economic value in industry [17] and can be classified into various classes, such as alkaloids, phenolics, saponins, and terpenes [18, 19].

Terpenoids, for example (tocotrienol, tocopherols, carotenoids, limonoids, and phytosterols), are already known antioxidants, as are some carotenoids [19, 20]. It is also noteworthy that some bioactive compounds from plants benefit health with anti-inflammatory, anticancer, antidiabetic, and digestive properties and favor blood circulation, which is the case of alkaloids [19, 21]. There are also studies investigating the use of bioactive compounds extracted from agricultural wastes to increase the ability to maintain and enhance food quality and health benefits [22]. These topics will be discussed later in this chapter.

Although economic production in Portugal has a strong support from agricultural practices and focused on forest species such as cork oaks (*Quercus suber*), holm oaks (*Quercus rotundifolia*), carob (*Ceratonia siliqua*), and maritime pine (*Pinus pinaster*), there are still limited data on valorization beyond traditional exploitation [23–27]. In the case of oak trees, apart from their long-standing use, these species contain a valuable compound known as suberin, the primary constituent of cork, which has a significant economic impact in Portugal. Suberin is also recognized in the health field for its bactericidal activity against *Escherichia coli* and *Staphylococcus aureus* [28–30].

Other examples of plant biomass valorization are gradually emerging, such as the use of agroforestry waste as raw material for biorefineries, the exploration of the bioactive potential of essential oil from the Portuguese endemic species *Santolina impressa* [31], and the use of phenolic compounds to enrich functional cookies, valorizing the residues from chestnut shell (*Castanea sativa*) [32] and arbutus (*Arbutus unedo*) pomace [33].

The variety of biomolecules found in Portugal's agroforestry systems presents a significant opportunity for sustainable development and the conservation of local species. However, it remains essential to encourage research and raise awareness about the economic and ecological importance of these biomolecules to ensure a sustainable future.

As previously mentioned, the extraction of such biomolecules may constitute a significant challenge as it involves not only selecting appropriate methods to ensure maximum yield and purity, but also addressing the inherent variability in plant materials and the need for sustainable practices. In the following section, these topics will be discussed in more detail.

3. Extraction methods for agroforestry biomolecules

Natural products (pure bioactive compounds or standardized plant extracts) from plant materials provide unlimited opportunities for different industrial applications [34, 35]. These compounds, known as secondary metabolites, are produced in response to a determined stress. Minor changes in the plant's physical characteristics due to biotic and abiotic factors can have a significant impact on the chemical profile and yield of compounds produced [36, 37]. Apart from the biotic and abiotic stresses, the selected extraction method is of major importance regarding its effect on the final constituents present in plant extracts and in its corresponding potential applicabilities [38].

These compounds can be extracted using different conventional and non-conventional methods. The most conventional techniques include Soxhlet, maceration, and percolation, among others. These methods are non-selective, typically require high temperatures, which are associated with an elevated energy cost, and often involve hazardous organic solvents (e.g., methanol, ethanol, propanol, ethyl acetate, acetone, hexane, etc.) [39]. However, they are commonly used as models to compare the efficiencies of alternative extraction methods [40].

Nowadays, the choice of more sustainable techniques and solvents is increasing. The creation, development, and use of chemical products and processes to minimize or completely remove the usage and production of dangerous substances is a broad definition of green chemistry [41]. As greener alternatives appear, other techniques have emerged throughout recent years, such as microwave-assisted extraction (MAE), ultrasound-assisted extraction (UAE), supercritical fluid extraction (SFE), subcritical water extraction (SWE), enzyme-assisted extraction (EAE), cold pressing extraction (CPE), high-pressure carbon dioxide assisted extraction (HPCDAE), among others [42]. These methods are generally renowned for their ability to almost entirely extract bioactive molecules from plant matrices, while using smaller amounts of organic solvents and requiring less energy and time [39]. In addition, more environmentally friendly solvents, such as natural deep eutectic solvents (NADES), ionic liquids (IL), supercritical fluids (SCF), and fluorous solvents (FS), have been widely proposed to replace conventional systems [43].

As mentioned before, despite their drawbacks, conventional techniques, such as Soxhlet [44–46], infusion [47–51], and maceration are still commonly used to extract distinct biomolecules. The latter case (maceration), which typically involves combining plant material with solvent in a vessel, assisted by heating and/or stirring, and leaving the mixture for a predetermined (usually long) time [43], is regarded as the most conventional method employed to extract phenolics from *Cistus ladanifer* [52], *Lavandula pedunculata* [47], *Melissa officinalis* [53], *Mentha pulegium* [51], *Origanum vulgare* [53], *Vaccinium myrtillus* [49, 54], *Vitis vinifera* [55, 56], *C. siliqua* pulp [50, 57, 58], and alkaloids from different species of *Calendula* [59] and *Lupinus* [60].

Among the newer and more environmentally friendly extraction techniques, UAE is the most widely applied method due to its modest requirements for solvent and energy. The ultrasonic waves employed in UAE result in cavitation forces that induce bubbles to burst and produce localized pressure in plant tissues. This pressure causes the rupture of plant cellular structure, which facilitates the diffusion of the solvent and increases the mass transport of intracellular components [43, 61]. Most of the studies regarding UAE still use organic solvents like acetone [62], ethanol [49, 62, 63], or methanol [64]. However, many recent studies consider the use of NADES as a replacement for traditional organic solvents for similar or even more effective and green extraction. NADES consist of a mixture of two or more components (hydrogen bond acceptor and donor), stabilized by strong intermolecular interactions, which have significantly lower melting points than any of their original constituents alone [65, 66]. Some reports have recognized these eco-friendly systems as new high-efficiency extraction solvents for polyphenols from the aromatic plants *L. pedunculata* [61], *Lavandula viridis* [67], and *Thymus lotocephalus* [67, 68] and residues (needles and bark) from maritime pine [69].

SWE is also a green extraction method that could selectively fractionate chemicals with distinct polarities from plant matrices by changing the solubility properties of water through temperature and pressure adjustments. More specifically, in a subcritical state, water can be used to extract molecules with medium to low polarity, similar to organic solvents [39]. The SWE has been found efficient for extracting several biomolecules from Portuguese forestry, namely, phenolic compounds like gallic and ellagic acids from *Q. suber* cork [70], as well as catechin and vanillic acid from blueberry pruning waste [54].

Other authors reported the extraction of phenolic acids (protocatechuic acid, pyrogallol, and gallic acid), flavonoids (quercetin-3-O-galactoside, catechin, procyanidin pentamer, and β-type proanthocyanidin tetramer) and alkaloids (chelidonine) from the aerial (stems, leaves, and fruits) and terrestrial (roots) parts of Portuguese plants, as well as from their residues, using other green alternative methods, such as MAE [54, 56, 71], HPCDAE [72, 73], and SFE [74].

Regardless of the inherent advantages and disadvantages of each extraction technique, the yield and efficiency of the extraction may be influenced by several factors. Among the most relevant parameters often studied are temperature, exposure time, solvent type, liquid-solid ratio, particle size, power, and pressure, all of which are typically optimized to determine the ideal extraction conditions for a given plant material and/or target biocompounds.

The largest class of plant secondary metabolites is the terpenoids. These compounds, formed from several 5-carbon entities called isoprene units (with additional functional groups), are found principally in essential oils (EOs). EOs are aromatic volatile liquids or semiliquids, mainly constituted of terpenes, terpenoids, and phenylpropanoids. They are typically isolated by distillation from the entire plant or various plant parts, such as flowers, leaves, fruits, stems, bark, peels, roots, or seeds [37]. Hydrodistillation (HD) and steam distillation (SD) are the main methods used in EO recovery and, according to the Pharmacopeia, the standard techniques for extracting EO for quality control [75]. However, solvent extraction can also be utilized for the fractionation of different aromatic volatile compounds from plants, including terpenoids [46, 59, 76]. In this context, "EO" refers to a distillation-derived product, whereas the term "extract" is related to the product resulting from other extraction processes (namely, those mentioned previously) [77]. SD, in which steam passes through the plant material with subsequent collection of volatile compounds, is used to isolate the EO from temperature-sensitive plants. In HD, the plant material is immersed in boiling water (ca. 3 h) where the volatile compounds are extracted by the steam and then condensed for their isolation. Due to the higher temperatures applied, HD has the disadvantages of possible hydrolysis of esters, polymerization of aldehydes, or decomposition of other thermosensitive compounds [75]. In this context, distinct methods will be beneficial for different EO isolation [75]. In fact, the quality and quantity of EO can be highly dependent on the isolation methods; it has been observed that for the EO of *Cupressus lusitanica* isolated by HD and SD, yields ranged between 0.12 and 0.26% (v/d.w.) for SD and 0.1% (v/d.w.) for HD [78]. Moreover, their major compounds also varied depending on the extraction method applied. However, recent research indicates HD as the most popular technique for isolating EO (even despite its disadvantages) [12, 79–82].

Several studies have reported 1,8-cineole (or eucalyptol), camphor, and/or α-pinene as the main abundant compounds of EO isolated by HD from *Lavandula* spp., namely, *L. pedunculata* [83, 84], *L. viridis* [85], and *L. stoechas* subsp. *luisieri* [86, 87]. In other species or subspecies of *Lavandula*, fenchone was the main chemotype in the EO profile of *L. pedunculata* [84], *L. pedunculata* subsp. *lusitanica* [86, 88], as well as the monoterpene carvacrol in the *Lavandula multifida* EO [89, 90]. The EO or extract of *Rosmarinus officinalis* is also very studied and dependent on the extraction technique, as has been evident in the different chemical profiles obtained by HD [91–93], simultaneous distillation-extraction [94], or HPCDAE with NADES.

During HD, an aqueous phase known as hydrolate arises, which is distilled with the EO and physically separated from it. These hydrolates are also good sources of bioactive compounds (especially phenol ethers, ketone alcohols, and methyl ethers) and, for that, also present numerous biological activities [95]. The hydrolate of *Aloysia citrodora*, whose major compounds are neral and geranial, showed excellent antimicrobial activity, inhibiting the growth of *E. coli* and *Candida albicans* in 90 and 80%, respectively [96]. In the case of the hydrolate of *Eucalyptus globulus*, 1,8-cineole was found the prevalent chemical component [96].

The solid residue of plant biomass and wastewater from HD can also have different purposes. For instance, solid residue can be used for power generation through direct combustion, composting, or as an absorbent, food supplement, or animal feed, while wastewater can have different biological activities [95]. Waste products from HD of *C. ladanifer* and *C. lusitanica* EO showed a great amount of biocompounds, such as α-pinene, camphene, *trans*-pinocarveol, limonene, sabinene, and δ-3-carene [78]. Due to the growing demand for EO in different industries, such as perfumery, cosmetics, aromatherapy, food and flavor, and pharmaceutics, researchers are looking for more innovative and cost-effective ways to generate high-yield and high-quality EO [75]. Their improvement using green energy sources, such as ultrasounds and microwaves, is particularly appealing from a sustainable and economic standpoint. Recently, the EO profile of the aromatic plant lemon verbena extracted with HD and microwave-assisted HD was compared and it was observed that the latter method achieved higher amounts of geranial and neral, as well as greater antioxidant and anti-tumor activities [81]. In addition to the mentioned greener techniques, another method has been widely used for the extraction of the main volatile or semi-volatile compounds without using solvents: solid phase microextraction (SPME). This technique allows the isolation of volatile compounds prior to their chromatographic analysis from different plant sources [97, 98], EOs or extracts [51], among others.

In the future, it is essential to continue combining research into plant chemistry and extraction technologies to lead toward advances that will improve yields and efficiency, reducing energy consumption while protecting the environment.

4. Characterization of extracted biomolecules

As previously described, the composition of agroforestry products is frequently challenging to determine due to the high concentration of macromolecules, such as sugars and minerals, which hinders the efficient analysis of minor components, including cellulose, lignin, and secondary metabolites, such as polyphenols. This section focuses on the main analytical techniques employed for the analysis of these molecules.

Spectroscopic techniques have been established as essential tools for biomolecule analysis, offering several notable advantages. These include being non-destructive, having a very short analysis time, being easy to use, and not requiring extensive sample preparation. Consequently, these techniques are often suitable to be operated by non-specialized users. Moreover, they provide quantitative data and can detect a wide range of compounds. In this context, colorimetric methodologies represent a significant area of analytical exploration. These methods often rely on the addition of a reagent that specifically reacts with the analyte to form a colored complex. The absorbance of this complex is then measured at a particular wavelength. By applying the Beer-Lambert law, a relationship is established between the measured absorbance and the concentration of the analyte in the sample.

The most employed colorimetric methods for the analysis of the composition of agroforestry products are the following: (i) The Folin-Ciocalteu Method, which employs a mixture of phosphomolybdic acid and phosphotungstic acid, known as Folin-Ciocalteu reagent, to reduce the phenolic compounds present in the sample. This reduction produces a blue-colored molybdenum-tungsten complex, whose absorbance is measured at 765 nm. To quantify the amount of total polyphenolic compounds total phenolic content (TPC), a calibration curve with gallic acid as the standard is typically employed. (ii) The aluminum chloride method involves the reaction of flavonoids present in a sample with chloride forming a yellow complex whose absorbance is measured at 415 nm. Quercetin is used as a standard for the quantification of total flavonoids content (FC). (iii) The vanillin-sulfuric acid method employs condensed tannins present in a sample to react with vanillin in acidic medium (concentrated sulfuric acid), producing a red complex. This reaction is specific for flavonoids containing catechin and epicatechin units. The absorbance is measured at 500 nm using catechin as a standard for the quantification of condensed tannin content (CT). For instance, Neiva et al. employed colorimetric methods for the quantification of TPC, FC, and CT in the bark of *E. globulus* and *Picea abies* (species with the highest prevalence in Portuguese industries) and *Acacia melanoxylon* and *Acacia dealbata* (the largest invasive species in Mediterranean countries, with an estimated area of ca. 16,000 ha in Portugal) [99]. Tavares et al. also employed colorimetric methods for the determination of TPC, FC, and CT in *C. lusitanica* and *C. ladanifer* wastes collected from northeastern Portugal [62].

Fourier transform infrared spectroscopy (FTIR) is also a valuable spectroscopic technique for providing additional information on the chemical bonds and functional groups present in agroforestry products, thus aiding in their identification and characterization. In this regard, Pedro et al. [100] have employed FTIR-ATR to characterize the phenolic compounds present in the floral extracts of *Acacia retinodes* and *Acacia mearnsii*, invasive species in Portugal. On the other hand, Magina et al. successfully used FTIR-ATR for the quantification of cellolignin present in the bark of *Eucalyptus globulus* collected in Portugal after enzymatic hydrolysis [101]. Additionally, Sartori et al. [102] employed FTIR analysis to examine the condensed tannins present in the barks of *Eucalyptus urophylla* hybrids originating from Portugal. Lastly, Barros et al. have successfully analyzed *P. pinaster* bark from Valença (northern Portugal) by FTIR for the quantification of lignin, cellulose, and hemicellulose [103].

Despite their simplicity, spectroscopic techniques also present some limitations. They are generally less selective and sensitive compared to more specific methods. Accurate results require prior calibration, and certain substances, such as sugars, can interfere at specific wavelengths. As a result, some researchers have opted to use alternative techniques for biomolecule analysis. In this regard, *chromatographic methods* offer significant advantages over scattering techniques, including high resolution and separation capacity, which results in superior sensitivity and specificity. These methods require the use of specialized detectors for the detection and quantification of specific analytes. Often, these methods are combined with mass spectrometry (MS), which has proven highly efficiency in identifying individual compounds in complex mixtures. In this regard, Sousa et al. [104] analyzed *Q. rotundifolia* bark collected in southern Portugal by gas chromatography-mass spectrometry (GC-MS) to quantify both graded acids and triterpenes present in the bark samples. In a recent study, Esteves et al. [105] investigated the chemical composition of young *Paulownia tomentosa* wood samples collected from a plantation in the Viseu region of Portugal using GC-MS to study the lipophilic composition. The samples were found to contain a variety of fatty acids, including palmitic, octadecadienoic, octadecenoic, stearic, and arachidic acids. Additionally, the samples were found to contain terpenic structures, including serruginol (a meroterpene), dehydroabietic acid (a resinic acid), β-sitosterol (a phytosterol), a lignan (sesamin), and alkanes such as heptacosane and pentacosane. Phenolic compounds, such as vanillin, vanillic acid, vanillic alcohol, and 4-hydroxy-3-methoxyphenylglycol, were also identified. In a recent study, Pedro et al. [106] identified a total of seven flavonoids in seven different pood samples of *Acacia* species (*A. melanoxylon, Acacia longifolia, Acacia cyclops, A. retinodes, Acacia pycnantha, A. mearnsii,* and *A. dealbata*) using liquid chromatography-electrospray ionization-mass spectrometry (LC-ESI-HRMS/MS). The identified flavonoids were myricetin, quercetin, kaempferol, rutin, and myricitrin, as well as the flavan-3-ol (+)-catechin and the flavanone naringenin. Moreover, the authors further employed high-pressure liquid chromatography coupled with a photodiode array detector (HPLC-DAD) to confirm the presence of phenolic compounds in the samples. Domingues et al. evaluated triterpenic compounds from the outer bark of different species of *Eucalyptus grandis* and *Eucalyptus urograndis* (*E. grandis* × *E. urophylla*) grown in Brazil and Portugal [46]. The extracts were analyzed by GC-MS to infer the fatty acids and long-chain aliphatic alcohols present in the samples.

It is often the case that data gathered from the analysis of biomolecules in complex matrices, such as those found in agroforestry products, originate large and intricate datasets that represent significant analytical challenges, making it difficult to draw clear conclusions. Therefore, the use of chemometric tools is often employed to extract relevant information. The most utilized statistical tool is the single-factor or multi-factor analysis of variance (ANOVA), which allows the assessment of the significant influence of various factors. As an illustrative example, the study conducted by Pedro et al. [100] employed a two-factor ANOVA to evaluate the differences between the flowers of *A. retinodes* and *A. mearnsii* as a function of species and flowering stage, utilizing the comprehensive chemical profiles of these species.

The unsupervised analysis methods, a cornerstone of chemometrics, are invaluable for uncovering intrinsic patterns in complex datasets without prior knowledge of sample classifications. Among the various techniques available, hierarchical cluster analysis (HCA) is particularly effective for grouping samples with similar multivariate characteristics. HCA operates by successively merging or splitting clusters, which can be visually represented through a dendrogram. The tree-like diagram provides a clear depiction of the hierarchical relationships between clusters, illustrating the nested arrangement of sample groups based on their similarity or dissimilarity metrics. Heatmaps provide an additional visual representation of the data matrix values, offering an intuitive and detailed visualization that complements the dendrogram. Each cell in a heatmap represents the intensity of a variable for a specific sample, with color gradients indicating the magnitude of these values. The combination of dendrograms and heatmaps in HCA enables a comprehensive understanding of the data structure, facilitating the identification of distinct sample groups and underlying patterns, which is particularly useful in different fields including metabolomics, genomics, and materials science.

In addition, principal component analysis (PCA) is a robust unsupervised data reduction technique that is widely employed in chemometrics to diminish the dimensionality of intricate datasets while maintaining their variance. By transforming the original variables into a new set of orthogonal components, PCA enables the extraction of the most significant features, which capture the underlying structure of the data. The principal components are linear combinations of the original variables, ordered according to the amount of variance they explain. The initial principal components frequently account for most of the total variance, thereby enabling researchers to visualize high-dimensional data in a lower-dimensional space, typically through scatter plots. This reduction in complexity facilitates the identification of patterns and relationships within the data, as well as the detection of outliers and clusters. PCA is particularly advantageous in domains, such as spectroscopy, chromatography, and environmental chemistry, where it enables the interpretation of intricate, multivariate data and enhances comprehension of the underlying chemical processes. For example, Pedro et al. [100, 106] employed the use of heatmaps and PCA for the analysis of *Acacia* flower samples according to their flowering stage or for the classification of *Acacia* pods varieties (*A. melanoxylon, A. longifolia, A. cyclops, A. retinodes, A. pycnantha, A. mearnsii,* and *A. dealbata*), based on their phenolic and flavonoid profile. This approach made it possible to distinguish the existing differences depending on the species and/or their flowering stage and the main compounds responsible for such differences. An example of the representation of a dendrogram as well as a PCA is shown in **Figure 1**.

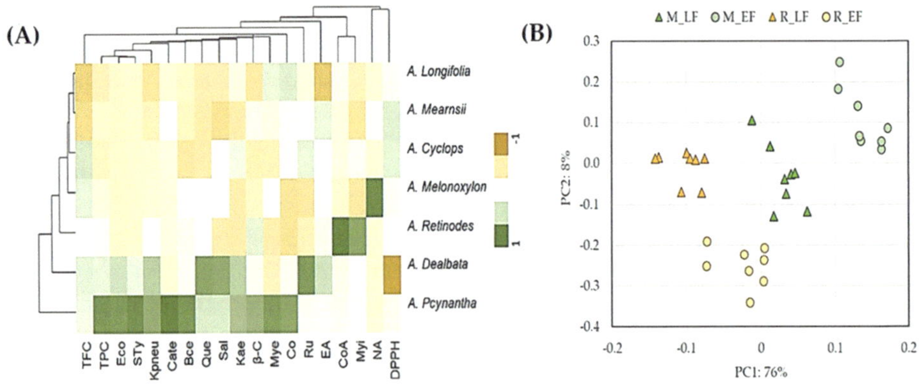

Figure 1.
(A) Heatmaps plotting clusters of content of phenol and flavonoid compounds of A. melanoxylon, A. longifolia, A. cyclops, A. retinodes, A. pycnantha, A. mearnsii, *and* A. dealbata. *(B) Score plot of PCA from* Acacia *flower samples analyzed.*

Linear discriminant analysis (LDA) is a supervised classification technique widely used in chemometrics to distinguish between different classes within a data set. Unlike unsupervised methods, LDA uses known class labels to find the linear combinations of features that best separate the classes. This is achieved by maximizing the ratio of between-class variance to within-class variance, ensuring that the projected data points are as distinct as possible while minimizing overlap. The resulting discriminant functions provide insight into the most significant variables contributing to class differentiation, facilitating understanding of the underlying factors that distinguish the classes. Borges et al. [107] employed chestnut orchards from the north of Portugal as a source of organic matter for soil management practices and successfully used LDA techniques to evaluate the significant differences according to the treatment and fractions employed in each soil.

5. Biotechnology tools for biomass valorization: Single-cell protein and biofuel production

Biotechnology offers cutting-edge solutions for transforming renewable biomass into high-value products like biofuels, bioplastics, and bioactive compounds. Using advanced techniques such as enzyme engineering, microbial fermentation, and metabolic pathway optimization, plant and agricultural residues can be efficiently converted into valuable chemicals. These innovative approaches not only boost the sustainability of industrial processes but also embrace circular economy principles by reducing waste and creating eco-friendly alternatives to traditional materials. A particularly exciting application is the production of single-cell proteins (SCP) from biomass, addressing the growing global demand for proteins [108]. This sustainable method optimizes waste management, allows for large-scale production with minimal environmental impact, and provides faster output without competing with conventional farming or being affected by climatic changes [109, 110].

SCP can be produced by various microorganisms, with fungi being the most used, followed by bacteria [111, 112]. SCP production is generally based on the transformation of biomass through a fermentation process, as illustrated in **Figure 2**.

Figure 2.
Schematic illustration of the SCP production process.

The composition of the SCP can vary depending on the methodology, substrate, and microorganisms used [113, 114]. Nutritionally, there is also considerable variation; for instance, the mycoprotein is effective in reducing total and LDL cholesterol, generally contains 30–50% protein, has low energy content and high fiber content, and is rich in micronutrients [115]. Its greatest advantage lies in its high levels of choline when compared to traditional protein sources (e.g., salmon, pork, soy and others) [116]. On the other hand, yeast SCP generally has 55–60% protein content, low lipid levels and high antioxidant properties [117, 118], while bacterial SCP is generally rich in proteins, carbohydrates, fats and fibers [119]. There are no exact values for SCP composition because, as mentioned, the nutrient content is strongly dependent on the substrate and the microorganisms used.

The implementation of SCP offers significant sustainability benefits, particularly in addressing global warming and land use challenges. However, significant research gaps remain [120]. Advancements in microbial engineering could enhance the competitiveness of SCP-based products regarding production costs, nutritional value, and functional properties, positioning them more favorably in the market [121]. Despite the various nutritional benefits, further studies are necessary, particularly on an industrial scale, since much of current knowledge is based on laboratory-scale research [122]. Increasing awareness and promoting the advantages of SCP as a dietary protein source is also crucial for broader consumer acceptance. To fully realize its potential, it is essential to intensify efforts in research and development, contributing not only to human nutrition but also to a more sustainable and resilient future [114]. Overall, the main challenges for SCP production lies in scaling up production, improving competitiveness, and ensuring consumer acceptance, particularly regarding food safety and sensory attributes [122, 123]. Although the concept of producing protein with minimal environmental impact is appealing, the SCP landscape remains underdeveloped. In Portugal, for example, there is only a brief report on a study involving the production of SCP using the fungus *Paecilomyces variotii* and spent sulfite liquor [124]. Attention and investment in advancing this technology remain limited, highlighting the need for stronger support to unlock SCP's full potential.

Energy consumption has increased worldwide, largely because of population growth, technological progress, and economic development. The reliance on non-renewable energy sources, primarily fossil fuels, significantly accelerates global warming, contributing to climate change, environmental degradation, and harmful effects on human health and biodiversity [125, 126]. To mitigate this, the European Union (EU) established a goal to improve energy efficiency by 32.5% until 2030 which included the incorporation of 20–30% of ethanol in gasoline [127]. In this regard, biofuels have emerged as a promising alternative to mitigate the harmful effects of fossil fuel consumption, reducing greenhouse gas emissions and dependence on non-renewable energy sources [125, 126, 128].

Biofuels can be produced from a wide range of biomass sources, including forestry byproducts, food crops, agricultural residues, and dedicated energy crops. Specifically, bioethanol can be produced from various renewable resources rich in carbohydrates, which can be hydrolyzed into fermentable sugars and subsequently converted into ethanol [126]. As previously discussed, mainland Portugal produces a significant amount of agroforestry-based wastes, including residues from the pulp and paper industry and agricultural residual biomass as cereal straw, corn stover, tree cutting and pruning residues, and pomaces as those from the olive oil production [125, 127, 129]. All these raw materials, usually considered as residues, are widely available at relatively low cost which, despite their different forms, are comparatively rich in lignocellulosic biomass. For example, *E. globulus* wood is the most used in the Portuguese pulp and paper industry due to its high cellulose content and low lignin content, which results in high pulping yields.

While lignin is not used for bioethanol production, it is a valuable source of high-value aromatic products [130]. The conversion of *E. globulus* bark into cellulosic sugars typically involves hydrolysis, preceded by a pretreatment where the kraft process is applied as a delignification step. Following this, alcoholic fermentation is carried out using ethanologenic microorganisms. Additionally, primary sludge from pulp and paper mills, which is rich in cellulose and hemicelluloses, can also be valorized for bioethanol production [131–134].

Due to the growing global interest in identifying new and sustainable carbohydrate sources for bioethanol production, alongside technological advancements, an increasing variety of feedstock materials has been researched. While finding cost-effective technologies for bioethanol production is a priority, the process remains significantly hindered by intrinsic biological challenges [135]. In the Mediterranean region, carob trees have been cultivated for centuries for their edible seed pods, yielding between 2000 and 3500 kg/ha, and are often used to diversify and revitalize dryland areas [136]. The seeds' endosperm is extracted to produce locust bean gum, a valuable natural food additive with strong gelling properties. The carob pod is typically used as animal feed or ground into powder for human consumption, but its high tannin content limits broader applications. Nevertheless, carob pods are rich in fermentable sugars like fructose, glucose, and sucrose, making them ideal for bioethanol production, a process that has already proven effective [137–140]. In addition, carob pulp has been selected for fermentative H_2 production using *Clostridium butyricum*, capable of producing 215.8 L H_2 per kilogram of dry weight. Although further studies are necessary to evaluate the feasibility and economic viability of scaling up the process, carob pods show great potential for the development of bioenergy production systems in Mediterranean regions [141].

6. Biomolecule valorization for agrofood

Water scarcity has intensified globally, particularly in arid and semi-arid regions. Projections indicate that by 2030, global water demand will rise by 50%, leading to critical shortages, further intensified by the impacts of climate change. With over 70% of global freshwater consumption allocated to agriculture, it is crucial to develop new solutions that enhance the efficiency of water and nutrient use

[142, 143]. In this context, absorbent polymer materials, such as hydrogels, have demonstrated significant potential for use in agriculture due to their high-water absorption and retention capacity. Environmentally friendly hydrogels can be synthesized from a variety of biomolecules, particularly natural polysaccharides, which can be sourced from aquatic environments, plants, bacteria, or animals [144]. The wide range of hydrogel applications in agriculture includes soilless cultivation, soil improvement, removal of heavy metals, smart materials, controlled release of fertilizers and agrochemicals, and water retention in soil [145–149]. Hydrogels formulated from biomolecules extracted from forestry or industrial residues have demonstrated biodegradability, non-toxicity, renewability, and cost-effectiveness compared to their synthetic counterparts. For instance, lignin-based hydrogels have been successfully tested in agricultural soils, showing potential for water retention, reducing nutrient leaching, and improving plant photosynthetic capacity under drought stress. Additionally, they have proven effective as soil conditioners [150, 151]. Natural gums, such as locust bean gum (LBG) extracted from carob seeds, have also been utilized in hydrogel formulations, exhibiting desirable properties, such as high swelling, biocompatibility, water retention, and biodegradability, making them popular not only in agricultural applications but also in the pharmaceutical, cosmetic, biomedical and food areas [136, 152–154].

As previously mentioned, the growing global population is putting increasing pressure on agricultural development to ensure a secure food supply. This has, in turn, led to a rise in pesticide use, which is recognized as a major contributor to environmental pollution. Current unsustainable practices must be replaced by innovative technologies capable of minimizing the use of harmful chemicals [155, 156]. Bio-based molecular agrochemicals have been developed with noteworthy advantages as high specificity and efficiency, biocompatibility and, more importantly, ecological sustainability [157]. Examples include antimicrobial peptides, that can be expressed by transgenic organisms or extracted from vegetable biomass and be topically applied [158–160]. As previously discussed, plant extracts contain a wide range of bioactive molecules and secondary metabolites, such as tannins, phenols, alkaloids, terpenoids, and flavonoids, all of which are regarded as highly appealing for the production of novel biopesticides [161–165]. In addition to their naturally occurring antibacterial and antioxidant properties, plant essential oils are also promising candidates for biopesticide development and have been successfully tested against fungi, viruses, and insects [85, 166, 167].

Plant extracts have garnered attention for their valuable potential in the food industry, utilizing locally sourced biomolecules from agricultural by-products and contributing to the development of sustainable agricultural practices. These biomolecules not only enhance the sensory attributes of food but also play a role in their preservation and offer health-promoting nutritional benefits. This has been recognized by Routray and Orsat, who discussed the significant amounts of waste and byproducts generated by agricultural practices and the food industry [168]. These byproducts are rich in bioactive components, including phenolic compounds, dietary fibers, amino acids, fatty acids and vitamins, among others. The Algarve region in Portugal could be a potential source of bioactive compounds, contributing to the creation of economically viable and environmentally friendly functional foods and beverages by reducing waste in the agri-food sector. Furthermore, the region's specific agricultural challenges, such as the long-standing impact of the Tomato yellow leaf curl virus and Tomato chlorosis virus on tomato production [169, 170], highlight the importance of ongoing research and surveillance in the agri-food sector. Such research is crucial for managing plant diseases and ensuring the quality and safety of agricultural products, thereby enabling their use in the food industry.

A wide variety of fruit constituents, such as seeds and peels, have been recovered from industrial waste processing to obtain value-added compounds, mostly phenolic, and incorporate them into food [43]. Carob bean gum, commonly known as LBG, is extracted from the fruit seeds/kernels endosperm of the carob tree, a plant species typical of the Algarve region but also abundant in other Mediterranean areas. LBG is a prominent example of a "Portuguese biomolecule" and is classified as a polysaccharide biopolymer from the group of galactomannans [136]. Due to its ability to provide high viscosity at low concentrations and act as a water binder, it is frequently used in various food products as a natural food additive (E410) with thickening and stabilizing functions. While LBG is well-known for its applications in the food industry, it has also demonstrated potential in the biopharmaceutical field, such as in the development of biodegradable non-toxic products and in drug delivery systems [171, 172]. Moreover, additional research has highlighted its use in treating specific health conditions such as hyperlipidemia and associated heart disease, diabetes, colon cancer, and gastrointestinal complications, further underscoring its medicinal value [171, 173, 174]. LBG also has numerous non-food applications, particularly in the textile, cosmetic, and construction industries [172].

Carob germ protein and oil are also notable carob-based systems recognized for their beneficial amino acid and fatty acid profiles, respectively, contributing to the nutritional enhancement of food products [175]. Additionally, carob pulp can be consumed directly or processed by roasting, grinding, and sieving to obtain flour or powder [176]. It has potential as a nutraceutical ingredient, serving as a healthier substitute for cocoa in food and beverage formulations due to its similar sensory properties and the absence of stimulating alkaloids like theobromine and caffeine [176]. Most of the commercial value of carob comes from the seed, where LBG plays a key role due to its wide range of applications, followed by the germ, which is a rich source of protein. However, the pulp is increasingly gaining economic importance as its nutritional benefits become more widely recognized [176, 177]. Carob pulp is high in bioactive compounds, including sugars, cyclitols, fibers, polyphenols, amino acids, vitamins, and minerals, while the seed mainly consists of three major components: gum (LBG), proteins (germ), and polyphenols [136, 173]. A schematic representation of the carob-derived products is shown in **Figure 3**.

Gum turpentine, derived from the steam distillation of oleoresin from various Mediterranean pine species, has applications in medicine as well as in the food, cosmetic, and detergent industries [178]. Additionally, wood extracts from *E. globulus*, which is abundant throughout Portugal, demonstrated the most promising results compared to almond shells, corn cobs, and grape pomace in inhibiting the growth of a wide range of food pathogenic bacteria in bioactive packaging [179]. Other relevant studies have reported that ethanol extracts from *A. unedo* (strawberry tree) fruits and *Ocimum basilicum* (basil) leaves, which are rich in catechin and rosmarinic acid, respectively, were successfully incorporated into bread loaves, conferring bioactive and preservative properties [180]. Additionally, extracts from *R. officinalis* have been shown to enrich the sensory attributes of bread [91].

As previously discussed, among plant extracts, EOs are recognized not only for their pleasant fragrances but also for their bioactive properties such as antioxidant, antimicrobial, anti-inflammatory and even anticancer activities. They are increasingly incorporated into food products for their health benefits and as natural preservatives [43, 181, 182]. In addition to the antioxidant and antimicrobial activities

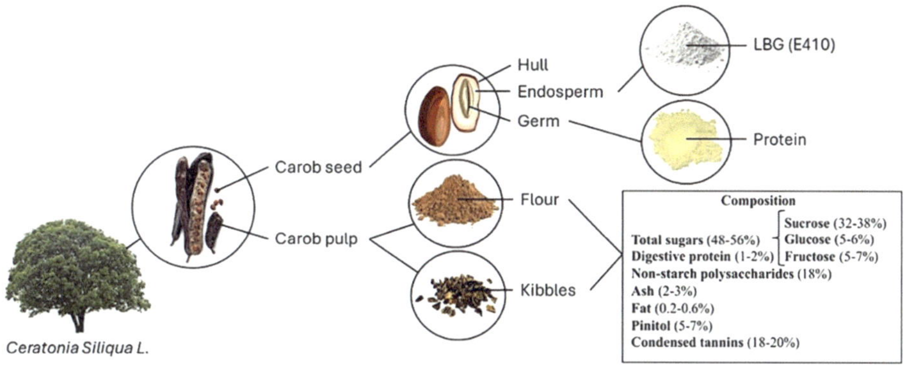

Figure 3.
Carob pod-derived food products and its pulp chemical composition (adapted from Solana et al. [176]).

reported in extracts from various pine species, including those found in Portugal [183–185], the EOs from *Pinus pinea*, *Pinus pinaster*, *Pinus sylvestris*, and *Pinus brutia* (highly predominant in Portugal and in the Mediterranean area) contain compounds regularly used in food products for flavoring and aroma, such as limonene, thymol and caryophyllene [178, 186–188]. Furthermore, EOs from Portuguese species, such as thyme, oregano, rosemary, verbena, basil, peppermint, pennyroyal and mint-beyond their traditional culinary uses-have shown significant inhibitory effects against various foodborne pathogens. This suggests their potential utility in enhancing food safety and extending shelf life [189]. Additionally, *Coriandrum sativum* and *Foeniculum vulgare* are two aromatic and medicinal species from the Apiaceae family, whose extracted EOs are noteworthy for applications not only in food but also in the pharmaceutical and cosmetic industries. The chemical compositions of EOs from these species, such as linalool in coriander and *trans*-anethole in fennel, are crucial for their potential applications [190]. Moreover, the diversity of leaf trichomes in *Lavandula* species from Portugal, which produce EOs, may contribute to the variety and efficacy of these oils [191]. *L. stoechas* subsp. *luisieri* and *L. pedunculata* EOs are two examples of Portuguese endemic *Lavandula* species that have recently been shown antibacterial and antifungal effects [83]. Interestingly, while the EO of *Laurus nobilis* is less studied, it has been identified as a potential ingredient for flavoring and preserving Mediterranean food products, including olive oil [192]. On the other hand, the EO from *Anethum graveolens*, known as dill, has attracted considerable research attention for its culinary applications, particularly due to its antimicrobial, antifungal, and antioxidant effectiveness [193]. Similarly, the wild *Smyrnium olusatrum* has demonstrated inhibitory effects against a wide range of fungi, especially *Cryptococcus neoformans*, suggesting their potential applications in food preservation [194]. The development of nanoemulsions and biodegradable films incorporating EOs for food packaging further highlights innovative approaches to utilizing these oils in the food industry [195].

Given that certain Portuguese plants belong to a broader group of botanicals with chemical constituents capable of enhancing the food industry through their aromatic and preservative qualities, further research, development, and commercialization are essential to fully leverage the potential of these biomolecules in food applications. This highlights the importance of effectively understanding and utilizing these natural resources.

7. Progress in biomolecule functionalization

As discussed in previous sections, biomolecules can be efficiently extracted from various Portuguese agroforestry residues, characterized by advanced methods and utilized in a wide range of applications. However, to enhance their properties and broaden their applications in fields such as materials science, biotechnology, and environmental science, strategic functionalization may be necessary. This can be accomplished through various chemical modifications, employing greener methods to ensure sustainability and environmental friendliness.

It was already discussed that the major components of woody and agricultural biomasses are cellulose, hemicellulose, and lignin. These three primary compounds, along with tannins, contain high levels of hydroxyl groups of varying natures, which confer them significant potential for surface chemistry modification. Additionally, the presence of aromatic moieties in lignin and tannins provides further sites for functionalization. In the literature, several functionalization procedures have been reported for bio-based polymers [196–199]. However, the functionalization procedures reported often employ chemicals which pose risks to human health and the environment, such as formaldehyde, thionyl chloride or sulfuryl chloride [200, 201]. It is thus necessary to develop of greener and safer functionalization procedures. In this context, mechanochemistry and esterification reactions using bio-based, environmentally friendly solvents and molecules (such as amino acids) have emerged as "greener" alternative strategies [202, 203]. The mechanochemical process, which combines mechanical forces, such as grinding or milling, with catalysis to enhance reaction rates and selectivity in the absence of solvents, is a relatively recent technique with significant potential for functionalizing bio-based molecules. Conversely, esterification reactions operate under mild conditions and employ simple catalysts, making them suitable for various types of modifications. Among the possible bio-based, safe and environmentally friendly solvents for esterification reactions, ethyl carbonate [204], γ-valerolactone [205], levulinic acid [206] and deep eutectic solvents [207–209] have emerged as suitable alternatives with great potential. Another important route for bio-based polymer functionalization is based on oxidation [210–212]. Several oxidant processes can be employed and among them, two should be highlighted due to their selectivity and versatility (**Figure 4**). First, the periodate-mediated oxidation produces a dialdehyde molecule in specific carbons of carbohydrates. Periodic acid and its salts, periodates, are known as regioselective oxidation agents capable of converting vicinal diols, such as carbohydrates, into dialdehyde structures. In the specific case of cellulose, the diol cleavage by periodate, under acidic conditions, at the C2 and C3 bond results in the formation of two aldehyde groups at the OH-C2 and OH-C3 positions, yielding dialdehyde cellulose (DAC) [213]. For example, Grenda et al. used this route to prepare different cationic and anionic cellulose derivatives with different applications [214]. The formed cellulose dialdehyde enables subsequent modifications using cationization agents such as Girard's reagent T (to form a stable imine structure with cationic quaternary ammoniums) [215] or anionization agents, such as sodium metabisulfite [210, 214]. Another important reaction is the oxidation mediated by 2,2,6,6-tetramethylpiperidine-1-oxyl radical (TEMPO). This reaction introduces carboxylic acid groups in carbohydrates, conferring charges to the polymers, at neutral or alkaline pH values. This reaction is widely used to oxidate the carbon C6 in cellulose, inducing swelling of the fibers and facilitating the preparation of nanocelluloses (nanofibers and nanocrystals) [212, 216].

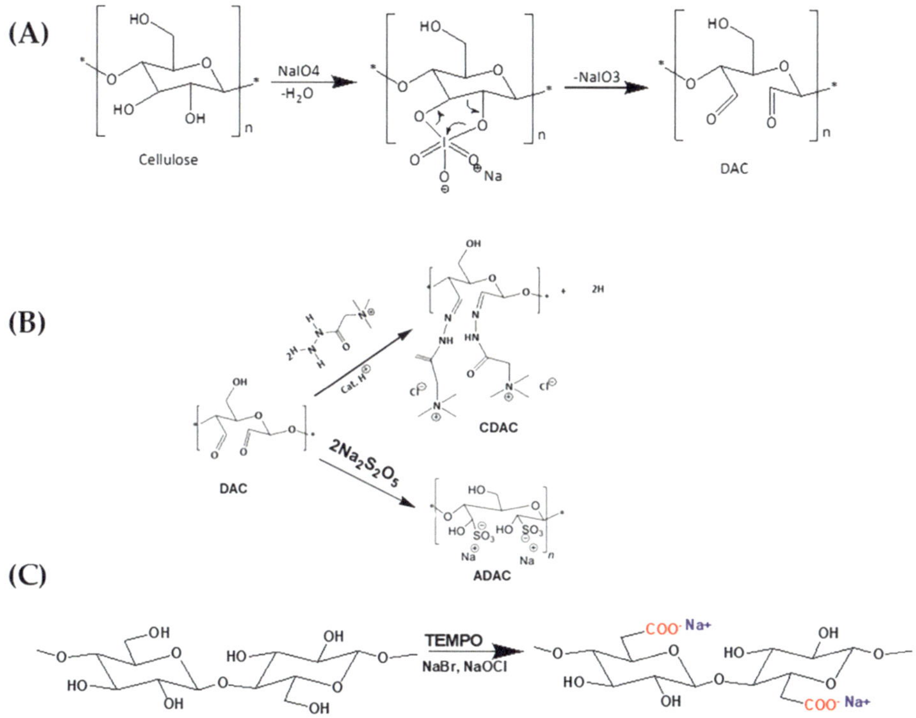

Figure 4.
(A) Reaction scheme of dialdehyde cellulose (DAC) synthesis; (B) cellulose dialdehyde cationization with Girard T reagent and anionization with sodium metabisulfite; (C) TEMPO mediated oxidation of cellulose.

Water-soluble cellulose derivatives can also be synthesized through etherification reactions. In these reactions, organic groups such as ethyl and methyl units react with the accessible hydroxyl groups of cellulose. This process yields cellulose ethers like carboxymethyl cellulose (CMC), hydroxypropyl methylcellulose, methylcellulose, hydroxyethyl cellulose, ethylcellulose, and hydroxypropyl cellulose (HPC) (**Figure 5**). Also, cationization by direct reaction of N-(3-chloro-2-hydroxypropyl) trimethylammonium chloride (CHPTAC) with OH groups of carbohydrates (C6 carbon) is achieved by the formation of an ether linkage, resulting in cationized cellulose. The reactive epoxy reagent is prepared in situ by reacting CHPTAC with an alkali. Anionic charges may be also added to cellulose by introducing, for instance, sulfonate groups ($-SO_3-$). One of the possibilities to prepare CMC is the direct sulfonation of cellulose in N,N-dimethylformamide using the $ClSO_3H$/DMF complex as a sulfonation agent, producing polyelectrolytes with good water solubility. Alternatively, NH_2SO_3H in DMF or sulfonation using sulfuric acid, SO_3, chlorosulfuric acid, and SO_2Cl_2, have also been used to obtain cellulose sulfate [217].

Conversely, cellulose esterification allows the preparation of other important cellulose derivatives, such as cellulose acetate and hydrophobically modified cellulose. Cellulose acetate is usually prepared by esterification of cellulose with acetic acid, substituting the hydroxyl groups of cellulose with acetyl groups [218]. On the other hand, sustainable hydrophobic modification of cellulose involves the use of renewable feedstocks, such as the use of fatty acids from plant oils. Through an esterification reaction, the fatty acids can be attached to the plenty hydroxyl groups present in carbohydrates (**Figure 5**). These modifications significantly enhance carbohydrate

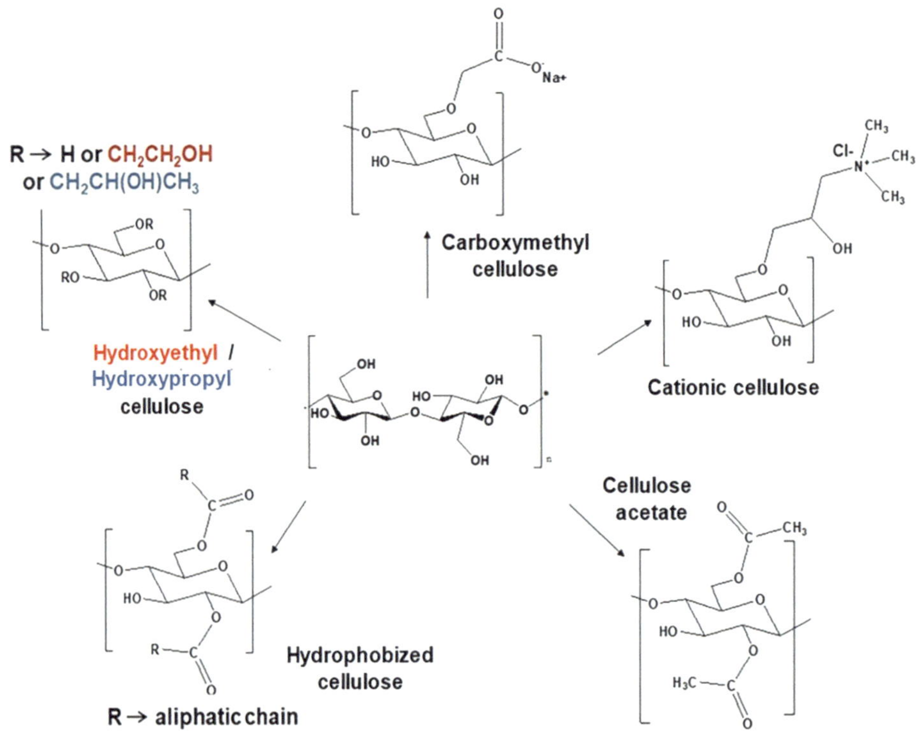

Figure 5.
Chemical structure of different cellulose derivatives, obtained through reactions in hydroxyl groups.

properties, enabling its use in a wide range of applications, from industrial materials to advanced biomedical devices, dispersants, and flocculants [219, 220].

The cationization and anionization procedures briefly described for cellulose have been also applied to other molecules, such as the previously discussed carob gum, LBG [221].

Lignin is another biomolecule that contains a variety of functional groups, including hydroxyls, methoxyls, carbonyls, and carboxyls and thus with high potential for functionalization. Such modifications often involve synthesizing new macromonomers that increase the reactivity of hydroxyl groups or alter other active chemical sites. Different chemical modifications, such as hydroxyalkylation, amination, nitration, sulfomethylation, and sulfonation, have been explored to introduce new functionalities into the lignin structure (**Figure 6**).

Hydroxyalkylation, amination, nitration, sulfomethylation and sulfonation of lignin introduce functional groups into the lignin structure, which can become ionizable and charged under different conditions [222]. Several chemical methods are used for this, with the Mannich reaction being one of the simplest. The main drawback is that formaldehyde is used in this reaction. Alternatively, cationization of lignin with CHPTAC can be attained by reaction with the phenolic hydroxyl groups of lignin, resulting in cationized lignin; similarly, phosphorylation of lignin is obtained by reacting the OH phenolic groups with phosphorus pentoxide, improving the fire-retardant properties of epoxy resins [223]. Tannins, due to their chemical similarity to lignin, can be also modified by employing similar chemical functionalization reactions used for lignin.

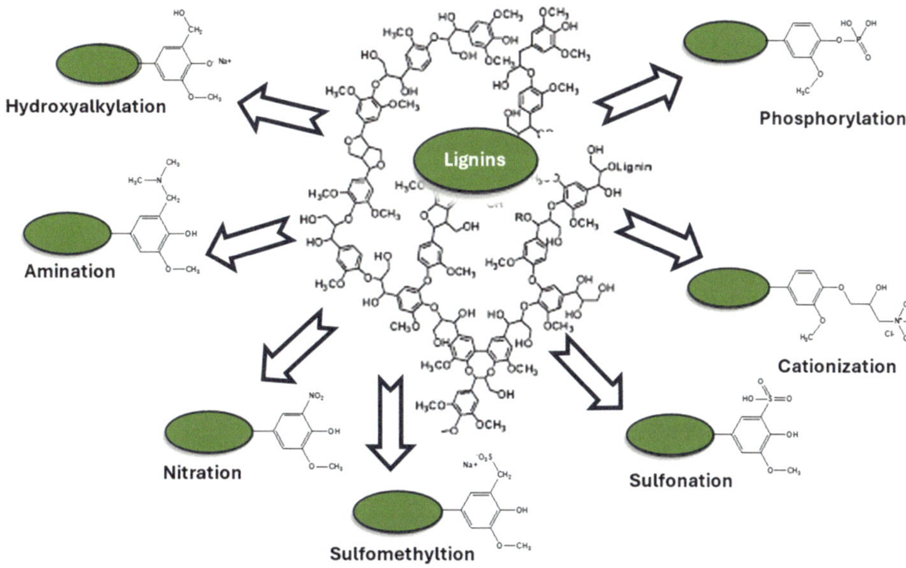

Figure 6.
Overview of chemical modifications of lignin.

The brief examples discussed above, clearly demonstrate that lignocellulosic molecules can be modified through a variety of approaches, each offering distinct enhancements to their properties. By integrating the principles of green chemistry and employing environmentally friendly solvents, these functionalization processes can become more sustainable and eco-conscious.

8. Advanced bio-based materials: The cellulose and lignin cases

Cellulose and its derivatives are increasingly being utilized to address contemporary challenges in biomedical, industrial, and environmental sectors. In the biomedical field, cellulose's inherent biocompatibility with human tissues, along with its capacity for chemical modification, make it an ideal candidate for applications, such as drug delivery, tissue engineering and wound healing [224–226]. Colloidal forms of cellulose, with their stable and homogeneous mixtures, offer innovative solutions in food, pharmaceutical, and cosmetic industries, enhancing product texture, consistency and stability. Moreover, the environmental applications of cellulose are considerable, ranging from water treatment and pollution control to the development of biodegradable packaging and sustainable textiles.

In the biomedical field, cellulose derivatives are highly attractive for controlled drug release systems [224, 225]. Numerous cellular species can be cultured on nanocellulose-based biomaterials, including hydrogels, nanofibers, sponges, composites, and membranes [227]. Among the different sources, bacterial nanocellulose is particularly suitable for cell culture due to its high purity, porosity, biodegradability, and low toxicity [228, 229]. Other cellulose derivatives have also been employed in biomedical applications. For instance, Bianchi et al. prepared hydrogels with wound dressing capabilities in association with β-cyclodextrin [230]. Suliwarno et al. [231] developed hydrogel-based materials formed by electron beam irradiation crosslinking of methylcellulose, also for wound dressing applications. Niemczyk-Soczynska et al.

[232] used methylcellulose to produce hydrogels via thermally induced crosslinking to create scaffolds for tissue engineering. Pasqui et al. [233] prepared carboxymethyl cellulose-hydroxyapatite hybrid hydrogels for bone tissue engineering and Wang et al. [234] successfully developed PEG–carboxymethyl cellulose nanoparticle hydrogels for injectable and thermosensitive drug delivery.

Beyond their biomedical applications, cellulose and its derivatives exhibit remarkable potential in other areas. In the food industry, cellulose has been successfully utilized as a stabilizer and thickener in products such as sauces, dressings, and dairy items [235]. It enhances texture, consistency, and mouthfeel, and helps to stabilize emulsions, preventing the separation of oil and water in products like salad dressings and beverages. Microcrystalline cellulose and CMC are common additives that improve the texture and consistency of these food products [236]. Additionally, colloidal cellulose can replace fat in low-calorie foods, mimicking the texture and mouthfeel of fat without compromising taste and consistency [237].

In the pharmaceutical and cosmetic industries, cellulose plays a crucial role in controlled-release formulations for drug delivery systems, ensuring a sustained and regulated release of active ingredients [238]. This enhances therapeutic efficacy and patient compliance. In cosmetics, cellulose acts as a thickening agent and stabilizer in creams, lotions, and gels, improving product consistency and stability [239, 240].

The paper and packaging industries also benefit from colloidal cellulose through its application in paper coatings, which improves the mechanical properties, printability, and surface smoothness [241]. Additionally, in industrial applications, cellulose serves as a rheology modifier in paints and coatings, enhancing their viscosity and application properties [242]. It stabilizes the dispersion of pigments and other components, ensuring a uniform and high-quality finish [243].

Additionally, cellulose and its derivatives offer significant advantages in environmental applications. For instance, in water treatment, cellulose derivatives are used for their ability to adsorb contaminants [214, 244]. Functionalized cellulose, such as modified cellulose nanocrystals, hydrogels, and cationic and anionic cellulose, can efficiently remove heavy metals, dyes, and organic pollutants from water [245]. This makes them effective adsorbents for purifying wastewater and ensuring clean drinking water in an environmentally friendly way.

In air and soil pollution control, cellulose-based materials are suitable to capture and neutralize harmful substances. For instance, cellulose aerogels can absorb airborne pollutants, including volatile organic compounds and small particles [246]. In soil remediation, cellulose derivatives help immobilize heavy metals and other toxic substances, preventing their leaching into groundwater and spreading contamination [247]. In agriculture, cellulose derivatives enhance soil health and water retention. Hydrogels made from cellulose can retain and slowly release water and nutrients to plants, improving crop yields and reducing irrigation needs [248]. Additionally, cellulose-based mulches and films protect soil from erosion and regulate soil temperature, promoting healthier plant growth [249].

Cellulose-based derivatives are also highly relevant in developing biodegradable packaging materials, which are essential for reducing plastic waste. These materials offer a sustainable alternative to conventional plastics, decomposing naturally without leaving harmful residues. Cellulose-based films, coatings, and laminates provide barrier properties against moisture, oxygen, and grease, extending the shelf life of food products while being environmentally friendly [250–253]. This supports the reduction of plastic pollution and promotes a circular economy.

In the textile industry, cellulose fibers are used to create sustainable and biodegradable fabrics. Lyocell, for example, is known for its strength, breathability, and eco-friendliness [254]. Produced through a closed-loop process that recycles solvents, lyocell minimizes environmental impact. Such cellulose-based textiles help reduce the dependence on synthetic fibers derived from non-renewable resources.

Another very interesting application is the development of flexible and lightweight electronic devices [255]. By incorporating cellulose nanofibers into substrates, researchers can create flexible displays and sensors [256]. There is also progress in making cellulose conductive by blending it with materials like carbon nanotubes or graphene, which could lead to the development of flexible, lightweight conductive films for various electronic uses [257]. Equally exciting is the use of cellulose in paper-based electronics [258]. By engineering cellulose-based paper, scientists can create low-cost, disposable devices such as sensors and RFID tags [256]. Furthermore, cellulose is being explored in energy storage devices, such as supercapacitors and batteries, due to its biodegradability and abundance, which offer a sustainable alternative to traditional materials [259]. Cellulose films have also been successfully developed to harvest energy via the development of novel triboelectric nanogenerators [260–262].

Similarly to cellulose, lignin acts as a plentiful renewable feedstock; as previously discussed, it is the primary natural source of aromatic monomers on Earth [130]. Although some efforts have been made to increase lignin's valorization, and its introduction into a wide range of markets has been proposed, it is still regarded as a by-product, being generally produced as a low value sub-product. This is the case of lignin from spent pulping liquors and chemical liberation of wood fibers from the pulp and paper industries, which are the main source of lignin for large scale use [263]. Most of the nearly 70 million tons of lignin annually produced by pulp and paper industries are wasted for heat generation [264], which clearly represents a waste of resources. Usually, lignin from pulp and paper is considered of low-quality due to its low molecular weight and the presence of contaminants, limiting its application. While some applications, for example biomedical materials, will require highly pure lignin extracts, low grade lignin can be applied in other cases such as the development of composite materials or flocculants for wastewater treatment. Lignin obtained from the paper industry could act, for example, as an inexpensive filler for composite materials to change its properties and reduce the production costs.

Lignin has some intrinsic properties that are valuable for a wide range of applications. This includes its antimicrobial activity, its high ability to absorb UV light, and its biocompatibility. The extraction of pure lignin could lead to valuable materials for wound dressing [265–267], cosmetics [268–270], and drug delivery systems [271, 272].

As previously mentioned, lignin contains several functional groups, including phenolic –OH, methoxy, and carbonyl groups, which provide excellent UV absorption, making it an appealing material for UV filtering [273]. Some studies report lignin's ability to improve the sun protection factor (SPF) of commercial sunscreens or moisturizing creams [274–278]. To overcome the potential marketing challenge associated with the dark color of lignin, some studies have been conducted to extract light-colored lignin in order to incorporate it in skin products [273, 279].

Light-colored lignin extracted from bamboo using choline chloride:glycerol:$AlCl_3$ DES was tested for its application as sunscreen agent [279]. Its sunscreen protection was determined by mixing it with pure cream or commercial SPF 15 sunscreen. While the pure cream only showed a SPF of 1.04, having almost no ability to absorb UV radiation, the addition of lignin significantly improved its absorption with the increasing lignin loading, reaching a SPF of 5.11 for 5 wt.% lignin addition. A similar trend was reported by Duy et al. [273] for sunscreens containing lignin extracted from coconut husk. The size of the lignin particles incorporated in the formulation also influences its ability to absorb UV light and confer sun protection; Girard et al. [280] showed that reducing the particle size from the macro- to the nanoscale increases the UV absorbance of a moisturizing cream containing similar lignin content.

In the biomedical field, lignin is considered an ecologically friendly and sustainable material with potential application for anti-aging, wound healing, biosensing, or drug delivery [281]. Its application is motivated by its antioxidant activity which results from its radical scavenging activity, and its excellent antimicrobial properties caused by the ability of the phenolic groups to damage and break bacterial cell membrane [282].

Lignin nano-systems can be used to encapsulate pharmaceuticals and allow for its targeted delivery to the desired cells by increasing the drugs stability to degradation. These delivery systems include lignin nanoparticles (LNPs), nano-capsules, and hydro- or nanogels, in which the active ingredients can be added by inclusion, adsorption, encapsulation, or chemical bonding [281]. The release of the drugs is driven by external stimuli, such as changes in the pH of the medium. Lu et al. showed the potential of LNPs to encapsulate ibuprofen (IBU) and increase its bioavailability and therapeutic efficacy [283]. LNPs had hydrophobic cores and hydrophilic shells, and IBU was encapsulated in the interior via hydrophobic interactions. The amphoteric nature of lignin contributed to the pH-sensitivity of the IBU-LNPs, making them stable at acidic conditions characteristic of the stomach (release of IBU $< 24\%$) and promoting the release of IBU at neutral conditions of the intestine (release of IBU up to 91%). In addition to reducing the irritation to the stomach, encapsulation of IBU in LPNs also decreased its cytotoxicity.

The use of lignin for biomedical use is also motivated by the fact that lignin can act simultaneously as a carrier and active ingredient. Reports show that lignin extracts have remarkable potential to prevent diseases caused by the overproduction of radicals, being a natural antioxidant and anti-cancer agents [284].

Lignin carriers were used to encapsulate other types of active ingredients, such as benzazulene [285], curcumin [286, 287], resveratrol [288, 289], doxorubicin [290, 291], among others. The potential of lignin as a carrier for drug delivery was recently reviewed by Chaudhary and Sinha [292].

Flocculation is an essential step in wastewater treatment. Flocculants promote particle aggregation and allow their removal by solid–liquid separation. Particle aggregation occurs by different mechanisms, which depend on the flocculant type, the nature of the material to be flocculated, and the conditions of the medium [293]. The use of lignin as a natural flocculant has been proposed as an alternative to non-sustainable agents currently used [294–298]. However, for an efficient flocculation performance, lignin is usually modified to improve its hydrophilicity and water solubility, following reactions, such as Mannich, sulfoalkylation, carboxylation, briefly described in subsection 7 [297]. Wang et al. proposed the production of lignin flocculants for dye removal from textile degumming black liquor waste [299]. At optimal conditions, dye removal ranged from 89 to 96%. Mannich-functionalized kraft lignin also showed remarkable results for flocculation of kaolin suspensions and three different dyes [300]. Varying the modification degree of the derivatives, its flocculation ability could be target for either removing kaolin or the dyes. Lignin derivative with low substitution and high solubility showed the optimal characteristics to remove 99.31% of kaolin by bridging mechanism, while highly charged and poorly soluble lignin derivative could remove more than 99.6% of dyes by charge attraction.

9. Challenges, prospects and conclusions

As demonstrated in this chapter, the extraction and valorization of biomolecules from agroforestry residues present several technical and economic challenges. On the technical side, the complexity of the biomolecule extraction processes is a key challenge. Biomolecules, such as polysaccharides, phenolics, and terpenes, are often embedded in complex plant matrices, requiring sophisticated methods for their isolation and purification. Conventional extraction techniques, like solvent extraction or steam distillation, may lead to inefficiencies or degradation of bioactive compounds, limiting their effectiveness. Advanced techniques, such as supercritical fluid extraction, enzymatic extraction, and membrane technologies, have emerged, offering more efficient and sustainable alternatives. However, these methods often demand significant capital investment in equipment, as well as expertise in optimizing parameters, such as pressure, temperature, and solvent composition, further hindering their industrial application.

Economically, the scalability of extraction technologies is still a significant barrier. While lab-scale extractions are feasible and often successful, the scaling-up transition implies challenges related to high operational costs and energy requirements. Moreover, the availability and quality of raw materials are often inconsistent, particularly in seasonal agroforestry feedstock, which can lead to fluctuations in production efficiency and cost-effectiveness. To achieve economic viability, extraction processes must be highly optimized and integrated into larger biorefinery operations that can process various feedstocks, maximize yield, and minimize waste. Investment in infrastructure, research, and development is critical to overcome these barriers and making biomolecule extraction from agroforestry residues commercially viable.

The commercialization of biomolecules is heavily influenced by national and international policy frameworks. In Portugal, as in many other countries, the regulatory landscape for bioproducts derived from agroforestry residues is evolving. The European Union's Green Deal and the Circular Economy Action Plan provide important incentives for the development of bio-based industries, aiming to reduce reliance on fossil-based resources. However, regulatory requirements concerning product safety, environmental impact, and intellectual property can be complex and difficult, particularly for small and medium-sized enterprises trying to enter the biomolecule market. For example, biomolecule-based products, such as nutraceuticals, cosmetics, or bio-based materials, must comply with rigorous regulations set by agencies, such as the European Food Safety Authority (EFSA) or the European Chemicals Agency (ECHA). These regulations often involve expensive and time-consuming testing to ensure the safety and efficacy of new bioproducts, which can delay market entry and raise costs. Additionally, the lack of harmonized standards for biomolecule extraction, characterization, and certification can further hinder commercialization efforts, as manufacturers face differing national regulations within the European market. To facilitate the growth of the bioeconomy, clearer, more supportive policy frameworks that streamline regulatory approval processes and incentivize innovation are needed.

Future research in biomolecule extraction and valorization is likely to focus on improving extraction efficiencies and expanding the range of high-value compounds that can be isolated from agroforestry residues. As discussed in this chapter, one emerging trend is the development of "green" extraction technologies, such as using ionic liquids or deep eutectic solvents, which are considered more environmentally friendly compared to traditional solvents. Research is also focusing on optimizing biorefinery processes that integrate various extraction methods to maximize the yield and purity of multiple biomolecules from a single biomass source, thus increasing the economic viability of such operations.

Another important research direction involves the valorization of agroforestry waste streams through the production of high-value bioproducts such as bioactive compounds for pharmaceuticals, nutraceuticals, and bio-based materials. Advances in biotechnology, including genetic engineering and metabolic pathway optimization, are enabling the production of customized biomolecules with enhanced functionalities. As discussed in this chapter, the use of machine learning and artificial intelligence to optimize extraction processes and predict yields is also gaining traction, offering new possibilities for reducing costs and increasing the efficiency of biomolecule production.

Emerging trends in circular economy models will likely play a key role in the future of biomolecule valorization. These models emphasize the reuse of agricultural and forestry waste, creating a more sustainable production cycle and reducing environmental impact. Collaboration between industries, research institutions, and policymakers will be essential in driving these innovations forward and ensuring their successful adoption in the marketplace.

Portugal has a rich agroforestry sector presenting vast potential for the extraction and valorization of biomolecules. The diverse ecosystems, including cork oak forests, olive groves, carob tree, vineyards, and pine forests, offer a wealth of renewable biomass resources that could be employed for high-value biomolecules. The Portuguese agroforestry sector is already a major contributor to the bioeconomy through the production of cork, olive oil, and wood products, but its potential for producing bio-based chemicals and materials remains largely unexplored. The key to unlocking this potential lies in investing in research and infrastructure that can support the scaling up of extraction technologies, as well as creating supportive regulatory frameworks that reduce barriers to commercialization. By doing so, Portugal could position itself as a leader in the bioeconomy, leveraging its abundant natural resources and innovative capabilities to produce sustainable, high-value biomolecules for global markets. The integration of advanced extraction technologies, coupled with clear policy support and continued research investment, will be crucial to unleash the full potential of Portugal's agroforestry sector for biomolecule extraction and valorization.

Acknowledgements

This work was funded by National Funds through FCT—Foundation for Science and Technology under the Projects "BIG opportunities in a small SEED: valorization of carob by-products into novel hydrogels for sustainable agriculture" (2022.07519. PTDC), UIDB/05183/2020 (https://doi.org/10.54499/UIDB/05183/2020; https://doi.org/10.54499/UIDP/05183/2020) and LA/P/0121/2020 (https://doi.org/10.54499/LA/P/0121/2020). Solange Magalhães, Catarina Fernandes and Inês Mansinhos acknowledge FCT for the PhD grants (https://doi.org/10.54499/2020.07638.BD; https://doi.org/10.54499/2021.05991.BD and https://doi.org/10.54499/SFRH/BD/145243/2019). Sandra Gonçalves acknowledges FCT for the financial support from CEECINST/00052/2021. Raquel Rodríguez Solana was supported by the grant RYC2022-036888-I, funded by MCIU/AEI/10.13039/501100011033 and by the FSE+.

Bruno Medronho and Luís Alves also acknowledge FCT for the individual research contracts CEECIND/01014/2018 (https://doi.org/10.54499/CEECIND/01014/2018/CP1540/CT0002) and 2021.00399.CEECIND ((https://doi.org/10.54499/2021.00399.CEECIND/CP1656/CT0025)), respectively. María José Aliano-González would like to thank the University of Cádiz for the grant in the modality "Margarita Salas", of the call for the Requalification of the Spanish University System for 2021–2023 (Resolution of the Rector of the University of Cadiz UCA/R155REC/2021, of 2 July), and financed by the Ministry of Universities of the Government of Spain through the European Recovery Instrument "Next Generation EU", of the European Union (Order UNI/551/2021, of 26May).

Author details

Bruno Medronho[1,2*], Hugo Duarte[1,3], Inês Mansinhos[1], João Brás[1], Ana Amorim[1], Isabela dos Anjos[1], Maria José Aliaño-González[1,4], Raquel Rodríguez-Solana[1,5], Luís Alves[3], Solange Magalhães[3], Catarina Fernandes[1,3], Sandra Gonçalves[1] and Anabela Romano[1]

1 Faculdade de Ciências e Tecnologia, MED–Mediterranean Institute for Agriculture, Environment and Development and CHANGE, Global Change and Sustainability Institute, Universidade do Algarve, Campus de Gambelas, Portugal

2 Surface and Colloid Engineering, FSCN Research Centre, Mid Sweden University, Sundsvall, Sweden

3 Department of Chemical Engineering, University of Coimbra, CERES, Coimbra, Portugal

4 Faculty of Sciences, Analytical Chemistry Department, University of Cádiz, Cádiz, Spain

5 Department of Agroindustry and Food Quality, Andalusian Institute of Agricultural and Fisheries Research and Training (IFAPA), Rancho de la Merced Center, Cádiz, Spain

*Address all correspondence to: bfmedronho@ualg.pt

References

[1] Gawel E, Pannicke N, Hagemann N. A path transition towards a bioeconomy—The crucial role of sustainability. Sustainability. 2019;**11**:3005

[2] Hsiao C-J, Hu J-L. Biomass and circular economy: Now and the future. Biomass. 2024;**4**:720-739

[3] Gericke M, Amaral AJR, Budtova T, De Wever P, Groth T, Heinze T, et al. The European polysaccharide network of excellence (EPNOE) research roadmap 2040: Advanced strategies for exploiting the vast potential of polysaccharides as renewable bioresources. Carbohydrate Polymers. 2024;**326**:121633. DOI: 10.1016/j.carbpol.2023.121633

[4] Nunes LJR, Meireles CIR, Pinto Gomes CJ, Almeida Ribeiro NMC. Historical development of the Portuguese forest: The introduction of invasive species. Forests. 2019;**10**:974

[5] Ferreira S, Monteiro E, Brito P, Vilarinho C. Biomass resources in Portugal: Current status and prospects. Renewable and Sustainable Energy Reviews. 2017;**78**:1221-1235. DOI: 10.1016/j.rser.2017.03.140

[6] Gaspar MC, Mendes CVT, Pinela SR, Moreira R, Carvalho MGVS, Quina MJ, et al. Assessment of agroforestry residues: Their potential within the biorefinery context. ACS Sustainable Chemistry & Engineering. 2019;**7**:17154-17165. DOI: 10.1021/acssuschemeng.9b03532

[7] Koul B, Yakoob M, Shah MP. Agricultural waste management strategies for environmental sustainability. Environmental Research. 2022;**206**:112285. DOI: 10.1016/j.envres.2021.112285

[8] Carmona-Cabello M, Garcia IL, Leiva-Candia D, Dorado MP. Valorization of food waste based on its composition through the concept of biorefinery. Current Opinion in Green and Sustainable Chemistry. 2018;**14**:67-79. DOI: 10.1016/j.cogsc.2018.06.011

[9] Resolution of the Council of Ministers. Resolução do Cons. Minist. n° 163/2017 Anexo 1-Plano Nac. para a Prooção das Biorrefinarias. Diário da República n.° 210/2017, Série I 2017-10-31, n.° 163/2017. Lisboa, Portugal: Imprensa Nacional-Casa da Moeda; 2017. pp. 5840-5847

[10] Braghiroli FL, Passarini L. Valorization of biomass residues from forest operations and wood manufacturing presents a wide range of sustainable and innovative possibilities. Current Forestry Reports. 2020;**6**:172-183. DOI: 10.1007/s40725-020-00112-9

[11] Bala S, Garg D, Sridhar K, Inbaraj BS, Singh R, Kamma S, et al. Transformation of agro-waste into value-added bioproducts and bioactive compounds: micro/nano formulations and application in the agri-food-pharma sector. Bioengineering. 2023;**10**:152. DOI: 10.3390/bioengineering10020152

[12] Pereira AJ, Porto M, Tauleigne-Gomes C. *Carduncellus Cuatrecasasii G.* López (Asteraceae) y *Eryngium Aquifolium* Cav. (Apiaceae), Dos Nuevas Especies Para La Flora de Portugal. Acta Botanica Malacitana. 2014;**39**:298-300. DOI: 10.24310/abm.v39i1.2590

[13] Carapeto A. Onobrychis Caput-Galli (L.) Lam., a new *Fabaceae* for the Portuguese flora. Acta Botanica

Malacitana. 2021;**46**:167-168. DOI: 10.24310/abm.v46i.13508

[14] Correia P, Barbosa C, Šimůnek Z, Muchagata J, Sá AA. A new species of Lesleya (Spermatopsida) from the carboniferous of Iberia and its palaeoecological and evolutionary significance. Historical Biology. 2023;**35**:185-196. DOI: 10.1080/08912963.2021.2025364

[15] Kvaček J, Mendes MM. A new species of the cheirolepidiaceous conifer pseudofrenelopsis from the lower cretaceous of Figueira Da Foz formation, Portugal. Review of Palaeobotany and Palynology. 2023;**309**:104821. DOI: 10.1016/j.revpalbo.2022.104821

[16] Mendes MM, Kvaček J, Doyle JA. *Pseudofrenelopsis Dinisii*, a new species of the extinct conifer family *Cheirolepidiaceae* from the probable lower Hauterivian (cretaceous) of Western Portugal. Review of Palaeobotany and Palynology. 2023;**315**:104905. DOI: 10.1016/j.revpalbo.2023.104905

[17] Kapadia P, Newell AS, Cunningham J, Roberts MR, Hardy JG. Extraction of high-value chemicals from plants for technical and medical applications. International Journal of Molecular Sciences. 2022;**23**:10334. DOI: 10.3390/ijms231810334

[18] Altemimi A, Lakhssassi N, Baharlouei A, Watson D, Lightfoot D. Phytochemicals: Extraction, isolation, and identification of bioactive compounds from plant extracts. Plants. 2017;**6**:42. DOI: 10.3390/plants6040042

[19] Jha AK, Sit N. Extraction of bioactive compounds from plant materials using combination of various novel methods: A review. Trends in Food Science and Technology. 2022;**119**:579-591. DOI: 10.1016/j.tifs.2021.11.019

[20] Dillard CJ, German JB. Phytochemicals: Nutraceuticals and human health. Journal of the Science of Food and Agriculture. 2000;**80**:1744-1756. DOI: 10.1002/1097-0010(20000915)80:12<1744::AID- JSFA725>3.0.CO;2-W

[21] Zhang J-J, Li Y, Zhou T, Xu D-P, Zhang P, Li S, et al. Bioactivities and health benefits of mushrooms mainly from China. Molecules. 2016;**21**:938. DOI: 10.3390/molecules21070938

[22] Parra-Pacheco B, Cruz-Moreno BA, Aguirre-Becerra H, García-Trejo JF, Feregrino-Pérez AA. Bioactive compounds from organic waste. Molecules. 2024;**29**:2243. DOI: 10.3390/molecules29102243

[23] Castro M, Teixeira A, Fernández-Núñez E. The nutritive value of different mediterranean browse species used as animal feeds under oak silvopastoral systems in northern Portugal. Agroforestry Systems. 2021;**95**:269-278. DOI: 10.1007/s10457-020-00588-1

[24] Sucena-Paiva L, Correia O, Rosário L, Chozas S. Holm oak wood pastures in SE Portugal: A spatial and temporal multiscale approach. Agroforestry Systems. 2022;**96**:173-186. DOI: 10.1007/s10457-021-00714-7

[25] Maia P, Rodrigues D, Carneiro S. Diversidade Biológica Em Povoamentos de *Pinus pinaster* Aiton Em Portugal–Reflexão Sobre Decisões de Gestão Aliadas à Conservação. Silva Lusitana. 2023;**31**:97-106. DOI: 10.1051/silu/20233102097

[26] Mahdad YM, Gaouar SBS. Origin, distribution and domestication of the carob tree (*Ceratonia Siliqua L.*). Turkish Journal of Botany. 2023;**47**:89-96. DOI: 10.55730/1300-008X.2748

[27] Alegria C, Almeida AM, Roque N, Fernandez P, Ribeiro MM. Species distribution modelling under climate change scenarios for maritime pine (*Pinus Pinaster* Aiton) in Portugal. Forests. 2023;**14**:591. DOI: 10.3390/f14030591

[28] Coelho IS, Pestana M. A Fileira Da Cortiça Em Portugal – Posicionamento e Competitividade. Silva Lusitana. 2021;**29**:73-100. DOI: 10.1051/silu/20212901073

[29] Rodrigues RML. Suberin Biotech Potential: From Bactericidal Nanoparticles to Optical Sensors [Thesis]. Portugal: Universidade NOVA de Lisboa; 2021

[30] Silva SP, Sabino MA, Fernandes EM, Correlo VM, Boesel LF, Reis RL. Cork: Properties, capabilities and applications. International Materials Review. 2005;**50**:345-365. DOI: 10.1179/174328005X41168

[31] Alves E, Simoes A, Domingues MR. Fruit seeds and their oils as promising sources of value-added lipids from agro-industrial byproducts: Oil content, lipid composition, lipid analysis, biological activity and potential biotechnological applications. Critical Reviews in Food Science and Nutrition. 2021;**61**:1305-1339. DOI: 10.1080/10408398.2020.1757617

[32] Pinto D, Moreira MM, Vieira EF, Švarc-Gajić J, Vallverdú-Queralt A, Brezo-Borjan T, et al. Development and characterization of functional cookies enriched with chestnut shells extract as source of bioactive phenolic compounds. Food. 2023;**12**:640. DOI: 10.3390/foods12030640

[33] Duarte H, Carrera C, Aliaño-González MJ, Gutiérrez-Escobar R, Jiménez-Hierro MJ, Palma M, et al. On the valorization of *Arbutus unedo* L. pomace: Polyphenol extraction and development of novel functional cookies. Food. 2023;**12**:3707. DOI: 10.3390/foods12193707

[34] Croteau R, Kutchan TM, Lewis NG. Natural products (secondary metabolites). In: Natural Products (Secondary Metabolites). Biochemistry and Molecular Biology of Plants. Rockville: American Society of Plant Physiologists; 2000. pp. 1250-1318

[35] Sasidharan S, Chen Y, Saravanan D, Sundram K, Latha L. Extraction, isolation and characterization of bioactive compounds from plants' extracts. African Journal of Traditional, Complementary and Alternative Medicines. 2010;**8**:1-10. DOI: 10.4314/ajtcam.v8i1.60483

[36] Khare S, Singh NB, Singh A, Hussain I, Niharika K, Yadav V, et al. Plant secondary metabolites synthesis and their regulations under biotic and abiotic constraints. Journal of Plant Biology. 2020;**63**:203-216. DOI: 10.1007/s12374-020-09245-7

[37] Mansinhos I, Gonçalves S, Romano A. How climate change-related abiotic factors affect the production of industrial valuable compounds in *Lamiaceae* plant species: A review. Frontiers in Plant Science. 2024;**15**:1370810-370837. DOI: 10.3389/fpls.2024.1370810

[38] Lefebvre T, Destandau E, Lesellier E. Selective extraction of bioactive compounds from plants using recent extraction techniques: A review. Journal of Chromatography. A. 2021;**1635**:461770. DOI: 10.1016/j.chroma.2020.461770

[39] Ferreira C, Moreira MM, Delerue-Matos C, Sarraguça M. Subcritical water

extraction to valorize grape biomass—A step closer to circular economy. Molecules. 2023;**28**:7538. DOI: 10.3390/molecules28227538

[40] Osorio-Tobón JF. Recent advances and comparisons of conventional and alternative extraction techniques of phenolic compounds. Journal of Food Science and Technology. 2020;**57**:4299-4315. DOI: 10.1007/s13197-020-04433-2

[41] Chemat F, Vian MA, Cravotto G. Green extraction of natural products: Concept and principles. International Journal of Molecular Sciences. 2012;**13**:8615-8627. DOI: 10.3390/ijms13078615

[42] Rahim MA, Ayub H, Sehrish A, Ambreen S, Khan FA, Itrat N, et al. Essential components from plant source oils: A review on extraction, detection, identification, and quantification. Molecules. 2023;**28**:6881. DOI: 10.3390/molecules28196881

[43] Leichtweis MG, Oliveira MBPP, Ferreira ICFR, Pereira C, Barros L. Sustainable recovery of preservative and bioactive compounds from food industry bioresidues. Antioxidants. 1827;**2021**:10. DOI: 10.3390/antiox10111827

[44] Custódio L, Escapa AL, Fernandes E, Fajardo A, Aligué R, Albericio F, et al. Phytochemical profile, antioxidant and cytotoxic activities of the carob tree (*Ceratonia siliqua* L.) germ flour extracts. Plant Foods for Human Nutrition. 2011;**66**:78-84. DOI: 10.1007/s11130-011-0214-8

[45] Custódio L, Fernandes E, Escapa AL, López-Avilés S, Fajardo A, Aligué R, et al. Antioxidant activity and *in vitro* inhibition of tumor cell growth by leaf extracts from the carob tree (*Ceratonia Siliqua*). Pharmaceutical Biology. 2009;**47**:721-728. DOI: 10.1080/13880200902936891

[46] Domingues RMA, Sousa GDA, Silva CM, Freire CSR, Silvestre AJD, Neto CP. High value triterpenic compounds from the outer barks of several *Eucalyptus* species cultivated in Brazil and in Portugal. Industrial Crops and Products. 2011;**33**:158-164. DOI: 10.1016/j.indcrop.2010.10.006

[47] Lopes CL, Pereira E, Soković M, Carvalho AM, Barata AM, Lopes V, et al. Phenolic composition and bioactivity of *Lavandula pedunculata* (mill.) Cav. Samples from different geographical origin. Molecules. 2018;**23**:1037. DOI: 10.3390/molecules23051037

[48] Pereira CG, Barreira L, da Rosa Neng N, Nogueira JMF, Marques C, Santos TF, et al. Searching for new sources of innovative products for the food industry within halophyte aromatic plants: *In vitro* antioxidant activity and phenolic and mineral contents of infusions and decoctions of *Crithmum maritimum* L. Food and Chemical Toxicology. 2017;**107**:581-589. DOI: 10.1016/j.fct.2017.04.018

[49] Plasencia P, Heleno SA, Finimundy T, Carocho M, Calhelha RC, Añibarro-Ortega M, et al. Recovery of high valuable bioactive molecules from *Vaccinium myrtillus* L. bioresidues. Waste and Biomass Valorization. 2023;**14**:2873-2884. DOI: 10.1007/s12649-023-02042-6

[50] Rodríguez-Solana R, Coelho N, Santos-Rufo A, Gonçalves S, Pérez-Santín E, Romano A. The influence of *in vitro* gastrointestinal digestion on the chemical composition and antioxidant and enzyme inhibitory capacities of carob liqueurs obtained with different elaboration techniques. Antioxidants. 2019;**8**:563. DOI: 10.3390/antiox8110563

[51] Rodríguez-Solana R, Esteves E, Mansinhos I, Gonçalves S, Pérez-Santín E, Galego L, et al. Influence of

elaboration process on chemical, biological, and sensory characteristics of European pennyroyal liqueurs. Journal of the Science of Food and Agriculture. 2021;**101**:4076-4089. DOI: 10.1002/jsfa.11043

[52] Barros L, Dueñas M, Alves CT, Silva S, Henriques M, Santos-Buelga C, et al. Antifungal activity and detailed chemical characterization of *Cistus ladanifer* phenolic extracts. Industrial Crops and Products. 2013;**41**:41-45. DOI: 10.1016/j.indcrop.2012.03.038

[53] Mascoloti Spréa R, Caleja C, Pinela J, Finimundy TC, Calhelha RC, Kostić M, et al. Comparative study on the phenolic composition and *in vitro* bioactivity of medicinal and aromatic plants from the *Lamiaceae* family. Food Research International. 2022;**161**:111875. DOI: 10.1016/j.foodres.2022.111875

[54] Dorosh O, Fernandes VC, Delerue-Matos C, Moreira MM. Blueberry pruning wastes: From an undervalued agricultural residue to a safe and valuable source of antioxidant compounds for the food industry. Food. 2024;**13**:317. DOI: 10.3390/foods13020317

[55] Duarte CN, Taofiq O, Dias MI, Heleno SA, Santos-Buelga C, Barros L, et al. Chemical characterization and bioactive properties of wine lees and diatomaceous earth towards the valorization of underexploited residues as potential cosmeceuticals. Cosmetics. 2023;**10**:58. DOI: 10.3390/cosmetics10020058

[56] Ueda JM, Griebler KR, Finimundy TC, Rodrigues DB, Veríssimo L, Pires TCSP, et al. Polyphenol composition by HPLC-DAD-(ESI-)MS/MS and bioactivities of extracts from grape agri-food wastes. Molecules. 2023;**28**:7368. DOI: 10.3390/molecules28217368

[57] Roseiro LB, Duarte LC, Oliveira DL, Roque R, Bernardo-Gil MG, Martins AI, et al. Supercritical, ultrasound and conventional extracts from carob (*Ceratonia siliqua* L.) biomass: Effect on the phenolic profile and antiproliferative activity. Industrial Crops and Products. 2013;**47**:132-138. DOI: 10.1016/j.indcrop.2013.02.026

[58] Vilas-Boas AM, Brassesco ME, Quintino AC, Vieira MC, Brandão TRS, Silva CLM, et al. Particle size effect of integral carob flour on bioaccessibility of bioactive compounds during simulated gastrointestinal digestion. Food. 2022;**11**:1272. DOI: 10.3390/foods11091272

[59] Faustino MV, Seca AML, Silveira P, Silva AMS, Pinto DCGA. Gas chromatography–mass spectrometry profile of four *Calendula L.* taxa: A comparative analysis. Industrial Crops and Products. 2017;**104**:91-98. DOI: 10.1016/j.indcrop.2017.04.029

[60] Valente IM, Sousa C, Almeida M, Miranda C, Pinheiro V, Garcia-Santos S, et al. Insights from the yield, protein production, and detailed alkaloid composition of white (*Lupinus albus*), narrow-leafed (*Lupinus angustifolius*), and yellow (*Lupinus luteus*) lupin cultivars in the Mediterranean region. Frontiers in Plant Science. 2023;**14**:1231777. DOI: 10.3389/fpls.2023.1231777

[61] Mansinhos I, Gonçalves S, Rodríguez-Solana R, Ordóñez-Díaz JL, Moreno-Rojas JM, Romano A. Ultrasonic-assisted extraction and natural deep eutectic solvents combination: A green strategy to improve the recovery of phenolic compounds from *Lavandula pedunculata* Subsp. *Lusitanica* (Chaytor) Franco. Antioxidants. 2021;**10**:582. DOI: 10.3390/antiox10040582

[62] Tavares CS, Martins A, Miguel MG, Carvalheiro F, Duarte LC, Gameiro JA,

et al. Bioproducts from Forest biomass II. Bioactive compounds from the steam-distillation by-products of *Cupressus lusitanica* mill. and *Cistus ladanifer* L. wastes. Industrial Crops and Products. 2020;**158**:112991. DOI: 10.1016/j.indcrop.2020.112991

[63] Mateus DMR, Ferraz E, Perna V, Sales P, Hipólito-Correia V. Essential oils and extracts of plants as biocides against microorganisms isolated from the ruins of the Roman city of Conímbriga (Portugal). Environmental Science and Pollution Research. 2023;**31**:40669-40677. DOI: 10.1007/s11356-023-28212-6

[64] Gonçalves S, Mansinhos I, Rodríguez-Solana R, Pereira-Caro G, Moreno-Rojas JM, Romano A. Impact of metallic nanoparticles on *in vitro* culture, phenolic profile and biological activity of two Mediterranean Lamiaceae species: Lavandula Viridis L'Hér and thymus Lotocephalus G. López and R. Morales. Molecules. 2021;**26**:6427. DOI: 10.3390/molecules26216427

[65] Paiva A, Craveiro R, Aroso I, Martins M, Reis RL, Duarte ARC. Natural deep eutectic solvents – solvents for the 21st century. ACS Sustainable Chemistry & Engineering. 2014;**2**:1063-1071. DOI: 10.1021/sc500096j

[66] Smith EL, Abbott AP, Ryder KS. Deep eutectic solvents (DESs) and their applications. Chemical Reviews. 2014;**114**:11060-11082. DOI: 10.1021/cr300162p

[67] Mansinhos I, Gonçalves S, Rodríguez-Solana R, Ordóñez-Díaz JL, Moreno-Rojas JM, Romano A. Impact of temperature on phenolic and osmolyte contents in *in vitro* cultures and micropropagated plants of two Mediterranean plant species, *Lavandula viridis* and *Thymus lotocephalus*. Plants. 2022;**11**:3516. DOI: 10.3390/plants11243516

[68] Mansinhos I, Gonçalves S, Rodríguez-Solana R, Duarte H, Ordóñez-Díaz JL, Moreno-Rojas JM, et al. Response of *Thymus lotocephalus in vitro* cultures to drought stress and role of green extracts in cosmetics. Antioxidants. 2022;**11**:1475. DOI: 10.3390/antiox11081475

[69] Duarte H, Gomes V, Aliano-Gonzalez MJ, Faleiro L, Romano A, Medronho B. Ultrasound-assisted extraction of polyphenols from maritime pine residues with deep eutectic solvents. Food. 2022;**11**:3754. DOI: 10.3390/foods11233754

[70] Cunha M, Lourenço A, Barreiros S, Paiva A, Simões P. Valorization of cork using subcritical water. Molecules. 2020;**25**:4695. DOI: 10.3390/molecules25204695

[71] Albuquerque BR, Prieto MA, Barreiro MF, Rodrigues A, Curran TP, Barros L, et al. Catechin-based extract optimization obtained from arbutus Unedo L. fruits using maceration/microwave/ultrasound extraction techniques. Industrial Crops and Products. 2017;**95**:404-415. DOI: 10.1016/j.indcrop.2016.10.050

[72] Alexandre AMRC, Matias A, Duarte CMM, Bronze MR. High-pressure CO_2 assisted extraction as a tool to increase phenolic content of strawberry-tree (*Arbutus unedo*) extracts. Journal of CO_2 Utilization. 2018;**27**:73-80. DOI: 10.1016/j.jcou.2018.07.002

[73] Gañán NA, Dias AMA, Bombaldi F, Zygadlo JA, Brignole EA, de Sousa HC, et al. Alkaloids from *Chelidonium majus* L.: Fractionated supercritical CO2 extraction with co-solvents. Separation and Purification Technology.

2016;**165**:199-207. DOI: 10.1016/j.seppur.2016.04.006

[74] Alexandre AMRC, Serra AT, Matias AA, Duarte CMM, Bronze MR. Supercritical fluid extraction of *Arbutus unedo* distillate residues – Impact of process conditions on antiproliferative response of extracts. Journal of CO2 Utilization. 2020;**37**:29-38. DOI: 10.1016/j.jcou.2019.11.002

[75] Pheko-Ofitlhile T, Makhzoum A. Impact of hydrodistillation and steam distillation on the yield and chemical composition of essential oils and their comparison with modern isolation techniques. Journal of Essential Oil Research. 2024;**36**:105-115. DOI: 10.1080/10412905.2024.2320350

[76] Vladić J, Kovačević S, Aladić K, Rebocho S, Jokić S, Podunavac-Kuzmanović S, et al. Novel insights into the recovery and stabilization of *Rosmarinus Officinalis* volatile aroma compounds using green solvents. Food and Bioprocess Technology. 2024;**17**:1215-1230. DOI: 10.1007/s11947-023-03188-w

[77] Perović AB, Karabegović IT, Krstić MS, Veličković AV, Avramović JM, Danilović BR, et al. Novel hydrodistillation and steam distillation methods of essential oil recovery from lavender: A comprehensive review. Industrial Crops and Products. 2024;**211**:118244. DOI: 10.1016/j.indcrop.2024.118244

[78] Tavares CS, Martins A, Faleiro ML, Miguel MG, Duarte LC, Gameiro JA, et al. Bioproducts from forest biomass: Essential oils and hydrolates from wastes of *Cupressus lusitanica* mill. and *Cistus ladanifer* L. Industrial Crops and Products. 2020;**144**:112034. DOI: 10.1016/j.indcrop.2019.112034

[79] Lima A, Arruda F, Janeiro A, Rodrigues T, Baptista J, Figueiredo AC, et al. Essential oils from different parts of *Azorean cryptomeria* Japonica (Thunb. Ex L.f.) D. Don (*Cupressaceae*): Comparison of the yields, chemical compositions, and biological properties. Applied Sciences. 2023;**13**:8375. DOI: 10.3390/app13148375

[80] Lima A, Arruda F, Wortham T, Janeiro A, Rodrigues T, Baptista J, et al. Chemical compositions and *in vitro* antioxidant activities of the essential oils of sawdust and resin-rich bark from *Azorean cryptomeria* Japonica (*Cupressaceae*). Antioxidants. 2024;**13**:728. DOI: 10.3390/antiox13060728

[81] Sprea RM, Fernandes LHM, Pires TCSP, Calhelha RC, Rodrigues PJ, Amaral JS. Volatile compounds and biological activity of the essential oil of *Aloysia citrodora* Paláu: Comparison of hydrodistillation and microwave-assisted hydrodistillation. Molecules. 2023;**28**:4528. DOI: 10.3390/molecules28114528

[82] Arruda F, Lima A, Wortham T, Janeiro A, Rodrigues T, Baptista J, et al. Sequential separation of essential oil components during hydrodistillation of fresh foliage from *Azorean cryptomeria* Japonica (*Cupressaceae*): Effects on antibacterial, antifungal, and free radical scavenging activities. Plants. 2024;**13**:1729. DOI: 10.3390/plants13131729

[83] Domingues J, Goulão M, Delgado F, Gonçalves JC, Gonçalves J, Pintado CS. Essential oils of two Portuguese endemic species of *Lavandula* as a source of antifungal and antibacterial agents. PRO. 2023;**11**:1165. DOI: 10.3390/pr11041165

[84] Zuzarte MR, Dinis AM, Cavaleiro C, Salgueiro LR, Canhoto JM. Trichomes, essential oils and *in vitro* propagation of *Lavandula pedunculata* (*Lamiaceae*).

Industrial Crops and Products. 2010;**32**:580-587. DOI: 10.1016/j.indcrop.2010.07.010

[85] Mansinhos I, Gonçalves S, Rodríguez-Solana R, Moreno-Rojas JM, Romano A. Environmental factors related to climate change alter the chemical composition and biological activity of *Lavandula viridis* L'Hér essential oil. Agriculture. 2024;**14**:1067. DOI: 10.3390/agriculture14071067

[86] Matos F, Miguel MG, Duarte J, Venâncio F, Moiteiro C, Correia AID, et al. Antioxidant capacity of the essential oils from *Lavandula luisieri*, *L. stoechas* Subsp. *Lusitanica*, *L. stoechas* Subsp. *Lusitanica* × *L. luisieri* and L. *viridis* grown in Algarve (Portugal). Journal of Essential Oil Research. 2009;**21**:327-336. DOI: 10.1080/10412905.2009.9700184

[87] Zuzarte M, Gonçalves MJ, Cruz MT, Cavaleiro C, Canhoto J, Vaz S, et al. *Lavandula luisieri* essential oil as a source of antifungal drugs. Food Chemistry. 2012;**135**:1505-1510. DOI: 10.1016/j.foodchem.2012.05.090

[88] Costa P, Gonçalves S, Valentão P, Andrade PB, Almeida C, Nogueira JMF, et al. Metabolic profile and biological activities of *Lavandula pedunculata* Subsp. *Lusitanica* (Chaytor) Franco: Studies on the essential oil and polar extracts. Food Chemistry. 2013;**141**:2501-2506. DOI: 10.1016/j.foodchem.2013.05.055

[89] Alves-Silva J, Zuzarte M, Cavaleiro C, Salgueiro L. Antibiofilm effect of *Lavandula multifida* essential oil: A new approach for chronic infections. Pharmaceutics. 2023;**15**:2142. DOI: 10.3390/pharmaceutics15082142

[90] Zuzarte M, Vale-Silva L, Gonçalves MJ, Cavaleiro C, Vaz S, Canhoto J, et al. Antifungal activity of phenolic-rich *Lavandula multifida* L. essential oil. European Journal of Clinical Microbiology & Infectious Diseases. 2012;**31**:1359-1366. DOI: 10.1007/s10096-011-1450-4

[91] Kessler JC, Vieira V, Martins IM, Manrique YA, Ferreira P, Calhelha RC, et al. Chemical and organoleptic properties of bread enriched with *Rosmarinus officinalis* L.: The potential of natural extracts obtained through green extraction methodologies as food ingredients. Food Chemistry. 2022;**384**:132514. DOI: 10.1016/j.foodchem.2022.132514

[92] Mata AT, Proença C, Ferreira AR, Serralheiro MLM, Nogueira JMF, Araújo MEM. Antioxidant and antiacetylcholinesterase activities of five plants used as Portuguese food spices. Food Chemistry. 2007;**103**:778-786. DOI: 10.1016/j.foodchem.2006.09.017

[93] Spréa RM, Caleja C, Finimundy TC, Calhelha RC, Pires TCSP, Amaral JS, et al. Chemical and bioactive evaluation of essential oils from edible and aromatic Mediterranean *Lamiaceae* plants. Molecules. 2024;**29**:2827. DOI: 10.3390/molecules29122827

[94] Ribeiro BS, de Ferreira MF, Moreira JL, Santos L. Simultaneous distillation–extraction of essential oils from *Rosmarinus officinalis* L. Cosmetics. 2021;**8**:117. DOI: 10.3390/cosmetics8040117

[95] Marcelino S, Hamdane S, Gaspar PD, Paço A. Sustainable agricultural practices for the production of medicinal and aromatic plants: Evidence and recommendations. Sustainability. 2023;**15**:14095. DOI: 10.3390/su151914095

[96] Almeida HHS, Crugeira PJL, Amaral JS, Rodrigues AE, Barreiro M-F.

Disclosing the potential of *Cupressus leylandii* A.B. Jacks & Dallim, *Eucalyptus globulus* Labill., *Aloysia citrodora* Paláu, and *Melissa officinalis* L. hydrosols as eco-friendly antimicrobial agents. Natural Products and Bioprospecting. 2024;**14**:1. DOI: 10.1007/s13659-023-00417-9

[97] Custódio L, Serra H, Nogueira JMF, Gonçalves S, Romano A. Analysis of the volatiles emitted by whole flowers and isolated flower organs of the carob tree using HS-SPME-GC/MS. Journal of Chemical Ecology. 2006;**32**:929-942. DOI: 10.1007/s10886-006-9044-9

[98] Gonçalves S, Serra H, Nogueira JMF, Almeida R, Custódio L, Romano A. Headspace-SPME of *in vitro* shoot-cultures and micropropagated plants of *Lavandula viridis*. Biologia Plantarum. 2008;**52**:133-136. DOI: 10.1007/s10535-008-0027-2

[99] Neiva DM, Luís Â, Gominho J, Domingues F, Duarte AP, Pereira H. Bark residues valorization potential regarding antioxidant and antimicrobial extracts. Wood Science and Technology. 2020;**54**:559-585. DOI: 10.1007/s00226-020-01168-3

[100] Pedro SI, Rosado T, Barroca C, Neiva D, Alonso-Herranz V, Gradillas A, et al. Characterisation of the phenolic profile of *Acacia retinodes* and *Acacia mearnsii* flowers' extracts. Plants. 2022;**11**:1442. DOI: 10.3390/plants11111442

[101] Magina S, Marques S, Gírio F, Lourenço A, Barros-Timmons A, Evtuguin DV. Chemical composition and structural features of cellolignin from steam explosion followed by enzymatic hydrolysis of *Eucalyptus globulus* bark. Industrial Crops and Products. 2024;**211**:118217. DOI: 10.1016/j.indcrop.2024.118217

[102] Sartori CJ, Mota GS, Miranda I, Mori FA, Pereira H. Tannin extraction and characterization of polar extracts from the barks of two *Eucalyptus urophylla* hybrids. BioResources. 2018;**13**:4820-4831. DOI: 10.15376/biores.13.3.4820-4831

[103] Barros D, Fernandes É, Jesus M, Barros L, Alonso-Esteban JI, Pires P, et al. The chemical characterisation of the maritime pine bark cultivated in Northern Portugal. Plants. 2023;**12**:3940. DOI: 10.3390/plants12233940

[104] Pinto PCRO, Sousa AF, Silvestre AJD, Neto CP, Gandini A, Eckerman C, et al. *Quercus suber* and *Betula pendula* outer barks as renewable sources of oleochemicals: A comparative study. Industrial Crops and Products. 2009;**29**:126-132. DOI: 10.1016/j.indcrop.2008.04.015

[105] Esteves B, Ferreira H, Viana H, Ferreira J, Domingos I, Cruz-Lopes L, et al. Termite resistance, chemical and mechanical characterization of *Paulownia tomentosa* wood before and after heat treatment. Forests. 2021;**12**:1114. DOI: 10.3390/f12081114

[106] Pedro SI, Fernandes TA, Luís Â, Antunes AMM, Gonçalves JC, Gominho J, et al. First chemical profile analysis of acacia pods. Plants. 2023;**12**:3486. DOI: 10.3390/plants12193486

[107] Borges O, Raimundo F, Coutinho J, Gonçalves B, Oliveira I, Martins A, et al. Carbon fractions as indicators of organic matter dynamics in chestnut orchards under different soil management practices. Agroforestry Systems. 2017;**92**:301-310. DOI: 10.1007/s10457-017-0088-3

[108] Bajpai P. Single Cell Protein Production from Lignocellulosic

Biomass. Springer Singapore: Singapore; 2017; Springer Briefs in Molecular Science. ISBN 978-981-10-5872-1

[109] Sharif M, Zafar MH, Aqib AI, Saeed M, Farag MR, Alagawany M. Single cell protein: Sources, mechanism of production, nutritional value and its uses in aquaculture nutrition. Aquaculture. 2021;**531**:735885. DOI: 10.1016/j.aquaculture.2020.735885

[110] Tropea A, Ferracane A, Albergamo A, Potortì AG, Lo Turco V, Di Bella G. Single cell protein production through multi food-waste substrate fermentation. Fermentation. 2022;**8**:91. DOI: 10.3390/fermentation8030091

[111] Onyeaka H, Anumudu CK, Okpe C, Okafor A, Ihenetu F, Miri T, et al. Single cell protein for foods and feeds: A review of trends. The Open Microbiology Journal. 2022;**16**:e187428582206160. DOI: 10.2174/1874 2858-v16-e2206160

[112] Rajput SD, Pandey N, Sahu K. A comprehensive report on valorization of waste to single cell protein: Strategies, challenges, and future prospects. Environmental Science and Pollution Research. 2024;**31**:26378-26414. DOI: 10.1007/s11356-024-33004-7

[113] Nadathur SR, Wanasundara JPD, Scanlin L. Chapter 1 - proteins in the diet: Challenges in feeding the global population. In: Nadathur SR, Wanasundara JPD, Scanlin LBT-SPS, editors. Sustainable Protein Sources. San Diego: Academic Press; 2017. pp. 1-19. ISBN 978-0-12-802778-3

[114] Bratosin BC, Darjan S, Vodnar DC. Single cell protein: A potential substitute in human and animal nutrition. Sustainability. 2021;**13**:9284. DOI: 10.3390/su13169284

[115] Derbyshire EJ, Delange J. Fungal protein – What is it and what is the health evidence? A systematic review focusing on mycoprotein. Frontiers in Sustainable Food Systems. 2021;**5**:581682. DOI: 10.3389/fsufs.2021.581682

[116] Wiedeman A, Barr S, Green T, Xu Z, Innis S, Kitts D. Dietary choline intake: Current state of knowledge across the life cycle. Nutrients. 2018;**10**:1513. DOI: 10.3390/nu10101513

[117] Jach ME, Serefko A, Ziaja M, Kieliszek M. Yeast protein as an easily accessible food source. Metabolites. 2022;**12**:63. DOI: 10.3390/metabo12010063

[118] Saravanan K, Rajendran M, Kathirvel P. Utilization of water hyacinth (*Eichhornia crassipes*) as a feasible substrate for the production of single cell protein using saccharomyces cerevisiae (Baker's yeast). Waste and Biomass Valorization. 2023;**14**:4231-4242. DOI: 10.1007/s12649-023-02138-z

[119] Kaur N, Chavan A. Production of single-cell protein from fruit and vegetable waste using an isolated strain of *Lactobacillus* Sp. Industrial Biotechnology. 2022;**18**:304-313. DOI: 10.1089/ind.2022.0023

[120] Aidoo R, Kwofie EM, Adewale P, Lam E, Ngadi M. Overview of single cell protein: Production pathway, sustainability outlook, and digital twin potentials. Trends in Food Science and Technology. 2023;**138**:577-598. DOI: 10.1016/j.tifs.2023.07.003

[121] Balagurunathan B, Ling H, Choi WJ, Chang MW. Potential use of microbial engineering in single-cell protein production. Current Opinion in Biotechnology. 2022;**76**:102740. DOI: 10.1016/j.copbio.2022.102740

[122] Salazar-López NJ, Barco-Mendoza GA, Zuñiga-Martínez BS,

Domínguez-Avila JA, Robles-Sánchez RM, Ochoa MAV, et al. Single-cell protein production as a strategy to reincorporate food waste and agro by-products back into the processing chain. Bioengineering. 2022;**9**:623. DOI: 10.3390/bioengineering9110623

[123] Koukoumaki DI, Tsouko E, Papanikolaou S, Ioannou Z, Diamantopoulou P, Sarris D. Recent advances in the production of single cell protein from renewable resources and applications. Carbon Resources Conversion. 2024;**7**:100195. DOI: 10.1016/j.crcon.2023.07.004

[124] Pereira SR, Ivanuša Š, Luzeiro J, Mateus C, Serafim LS, Xavier AMRB. Single cell protein production by *Paecilomyces variotii* from spent sulphite liquor. Journal of Biotechnology. 2010;**150**:509-509. DOI: 10.1016/j.jbiotec.2010.09.802

[125] Correia B, Matos HA, Lopes TF, Marques S, Gírio F. Sustainability assessment of 2G bioethanol production from residual lignocellulosic biomass. PRO. 2024;**12**:987. DOI: 10.3390/pr12050987

[126] Broda M, Yelle DJ, Serwańska K. Bioethanol production from lignocellulosic biomass—Challenges and solutions. Molecules. 2022;**27**:8717. DOI: 10.3390/molecules27248717

[127] Fontes AVP, João IM, Silva JM. Multicriteria evaluation of biomass residues in Portugal to second generation bioethanol production. Production. 2021;**31**:e20210060. DOI: 10.1590/0103-65 13.20210060

[128] Beluhan S, Mihajlovski K, Šantek B, Ivančić Šantek M. The production of bioethanol from lignocellulosic biomass: Pretreatment methods, fermentation, and downstream processing. Energies. 2023;**16**:7003. DOI: 10.3390/en16197003

[129] Lopes TF, Carvalheiro F, Duarte LC, Gírio F, Quintero JA, Aroca G. Techno-economic and life-cycle assessments of small-scale biorefineries for isobutene and xylo-oligosaccharides production: A comparative study in Portugal and Chile. Biofuels, Bioproducts and Biorefining. 2019;**13**:1321-1332. DOI: 10.1002/bbb.2036

[130] Melro E, Filipe A, Sousa D, Medronho B, Romano A. Revisiting lignin: A tour through its structural features, characterization methods and applications. New Journal of Chemistry. 2021;**45**:6986-7013. DOI: 10.1039/D0NJ06234K

[131] Amândio MST, Rocha JMS, Serafim LS, Xavier AMRB. Cellulosic bioethanol from industrial *Eucalyptus globulus* bark residues using Kraft pulping as a pretreatment. Energies. 2021;**14**:2185. DOI: 10.3390/en14082185

[132] Mendes CVT, Rocha JMS, Carvalho MGVS. Batch simultaneous saccharification and fermentation of primary sludge at very high solid concentrations for bioethanol production. Fermentation. 2023;**9**:888. DOI: 10.3390/fermentation9100888

[133] Seidl PR, Goulart AK. Pretreatment processes for lignocellulosic biomass conversion to biofuels and bioproducts. Current Opinion in Green and Sustainable Chemistry. 2016;**2**:48-53. DOI: 10.1016/j.cogsc.2016.09.003

[134] Branco RHR, Serafim LS, Xavier AMRB. Second generation bioethanol production: On the use of pulp and paper industry wastes as feedstock. Fermentation. 2018;**5**:4. DOI: 10.3390/fermentation5010004

[135] Tse TJ, Wiens DJ, Reaney MJT. Production of bioethanol—A review of factors affecting ethanol yield. Fermentation. 2021;7:268. DOI: 10.3390/fermentation7040268

[136] Martins-Loução MA, Correia PJ, Romano A. Carob: A Mediterranean resource for the future. Plants. 2024;**13**:1188. DOI: 10.3390/plants13091188

[137] Sánchez S, Lozano LJ, Godínez C, Juan D, Pérez A, Hernández FJ. Carob pod as a feedstock for the production of bioethanol in Mediterranean areas. Applied Energy. 2010;**87**:3417-3424. DOI: 10.1016/j.apenergy.2010.06.004

[138] Karimi F, Mazaheri D, Saei Moghaddam M, Mataei Moghaddam A, Sanati AL, Orooji Y. Solid-state fermentation as an alternative technology for cost-effective production of bioethanol as useful renewable energy: A review. Biomass Conversion and Biorefinery. 2021. DOI: 10.1007/s13399-021-01875-2

[139] Bahry H, Pons A, Abdallah R, Pierre G, Delattre C, Fayad N, et al. Valorization of carob waste: Definition of a second-generation bioethanol production process. Bioresource Technology. 2017;**235**:25-34. DOI: 10.1016/j.biortech.2017.03.056

[140] Lima-Costa ME, Tavares C, Raposo S, Rodrigues B, Peinado JM. Kinetics of sugars consumption and ethanol inhibition in carob pulp fermentation by *Saccharomyces cerevisiae* in batch and fed-batch cultures. Journal of Industrial Microbiology & Biotechnology. 2012;**39**:789-797. DOI: 10.1007/s10295-011-1079-4

[141] Ortigueira J, Silva C, Moura P. Assessment of the adequacy of different Mediterranean waste biomass types for fermentative hydrogen production and the particular advantage of carob (*Ceratonia siliqua* L.) pulp. International Journal of Hydrogen Energy. 2018;**43**:7773-7783. DOI: 10.1016/j.ijhydene.2018.03.024

[142] Oladosu Y, Rafii MY, Arolu F, Chukwu SC, Salisu MA, Fagbohun IK, et al. Superabsorbent polymer hydrogels for sustainable agriculture: A review. Horticulturae. 2022;**8**:605. DOI: 10.3390/horticulturae8070605

[143] Khan F, Atif M, Haseen M, Kamal S, Khan MS, Shahid S, et al. Synthesis, classification and properties of hydrogels: Their applications in drug delivery and agriculture. Journal of Materials Chemistry B. 2022;**10**:170-203. DOI: 10.1039/D1TB01345A

[144] Rial-Hermida MI, Rey-Rico A, Blanco-Fernandez B, Carballo-Pedrares N, Byrne EM, Mano JF. Recent progress on polysaccharide-based hydrogels for controlled delivery of therapeutic biomolecules. ACS Biomaterials Science & Engineering. 2021;7:4102-4127. DOI: 10.1021/acsbiomaterials.0c01784

[145] Qin C, Wang H, Zhao Y, Qi Y, Wu N, Zhang S, et al. Recent advances of hydrogel in agriculture: Synthesis, mechanism, properties and applications. European Polymer Journal. 2024;**219**:113376. DOI: 10.1016/j.eurpolymj.2024.113376

[146] Liu Y, Wang J, Chen H, Cheng D. Environmentally friendly hydrogel: A review of classification, preparation and application in agriculture. Science of the Total Environment. 2022;**846**:157303. DOI: 10.1016/j.scitotenv.2022.157303

[147] Zhang H, Yang M, Luan Q, Tang H, Huang F, Xiang X, et al. Cellulose anionic

hydrogels based on cellulose nanofibers as natural stimulants for seed germination and seedling growth. Journal of Agricultural and Food Chemistry. 2017;**65**:3785-3791. DOI: 10.1021/acs.jafc.6b05815

[148] Dhiman A, Thaper P, Bhardwaj D, Agrawal G. Biodegradable dextrin-based microgels for slow release of dual fertilizers for sustainable agriculture. ACS Applied Materials & Interfaces. 2024;**16**:11860-11871. DOI: 10.1021/acsami.3c16670

[149] Hsu HH, Zhang X, Xu K, Wang Y, Wang Q, Luo G, et al. Self-powered and plant-wearable hydrogel as LED power supply and sensor for promoting and monitoring plant growth in smart farming. Chemical Engineering Journal. 2021;**422**:129499. DOI: 10.1016/j.cej.2021.129499

[150] Mazloom N, Khorassani R, Zohuri GH, Emami H, Whalen J. Development and characterization of lignin-based hydrogel for use in agricultural soils: Preliminary evidence. CLEAN – Soil, Air, Water. 2019;**47**:1900101. DOI: 10.1002/clen.201900101

[151] Song B, Liang H, Sun R, Peng P, Jiang Y, She D. Hydrogel synthesis based on lignin/sodium alginate and application in agriculture. International Journal of Biological Macromolecules. 2020;**144**:219-230. DOI: 10.1016/j.ijbiomac.2019.12.082

[152] Başyiğit B, Altun G, Yücetepe M, Karaaslan A, Karaaslan M. Locust bean gum provides excellent mechanical and release attributes to soy protein-based natural hydrogels. International Journal of Biological Macromolecules. 2023;**231**:123352. DOI: 10.1016/j.ijbiomac.2023.123352

[153] Hayrabolulu H, Şen M, Çelik G, Kavaklı PA. Synthesis of carboxylated locust bean gum hydrogels by ionizing radiation. Radiation Physics and Chemistry. 2014;**94**:240-244. DOI: 10.1016/j.radphyschem.2013.05.048

[154] Mandal S, Chi H, Moss RE, Dhital P, Babatunde EO, Gurav R, et al. Seed gum-based polysaccharides hydrogels for sustainable agriculture: A review. International Journal of Biological Macromolecules. 2024;**263**:130339. DOI: 10.1016/j.ijbiomac.2024.130339

[155] Schmaltz S, Silva MA, Ninaus RG, Guedes JVC, Zabot GL, Tres MV, et al. Biomolecules in modern and sustainable agriculture. 3 Biotech. 2023;**13**:70. DOI: 10.1007/s13205-023-03486-2

[156] Kumar S, Bhanjana G, Sharma A, Dilbaghi N, Sidhu MC, Kim K-H. Development of nanoformulation approaches for the control of weeds. Science of the Total Environment. 2017;**586**:1272-1278. DOI: 10.1016/j.scitotenv.2017.02.138

[157] Li P, Huang Y, Fu C, Jiang SX, Peng W, Jia Y, et al. Eco-friendly biomolecule-nanomaterial hybrids as next-generation agrochemicals for topical delivery. EcoMat. 2021;**3**:e12132. DOI: 10.1002/eom2.12132

[158] Topman S, Tamir-Ariel D, Bochnic-Tamir H, Stern Bauer T, Shafir S, Burdman S, et al. Random peptide mixtures as new crop protection agents. Microbial Biotechnology. 2018;**11**:1027-1036. DOI: 10.1111/1751-7915.13258

[159] Montesinos E. Antimicrobial peptides and plant disease control. FEMS Microbiology Letters. 2007;**270**:1-11. DOI: 10.1111/j.1574-6968.2007.00683.x

[160] Zheng S, Jiang J, He M, Zou S, Wang C. Effect of kelp waste extracts on

the growth and development of Pakchoi (*Brassica chinensis* L.). Scientific Reports. 2016;**6**:38683. DOI: 10.1038/srep38683

[161] Lyubenova A, Rusanova M, Nikolova M, Slavov SB. Plant extracts and *Trichoderma* Spp: Possibilities for implementation in agriculture as biopesticides. Biotechnology and Biotechnological Equipment. 2023;**37**:159-166. DOI: 10.1080/13102818.2023.2166869

[162] Lopes AIF, Monteiro M, Araújo ARL, Rodrigues ARO, Castanheira EMS, Pereira DM, et al. Cytotoxic plant extracts towards insect cells: Bioactivity and nanoencapsulation studies for application as biopesticides. Molecules. 2020;**25**:5855. DOI: 10.3390/molecules25245855

[163] Kumar J, Ramlal A, Mallick D, Mishra V. An overview of some biopesticides and their importance in plant protection for commercial acceptance. Plants. 2021;**10**:1185. DOI: 10.3390/plants10061185

[164] Fragkouli R, Antonopoulou M, Asimakis E, Spyrou A, Kosma C, Zotos A, et al. Mediterranean plants as potential source of biopesticides: An overview of current research and future trends. Metabolites. 2023;**13**:967. DOI: 10.3390/metabo13090967

[165] Saad MMG, Abdelgaleil SAM, Shiono Y. Antibacterial and herbicidal properties of secondary metabolites from fungi. Natural Product Research. 2021;**35**:5446-5451. DOI: 10.1080/14786419.2020.1779718

[166] Khursheed A, Rather MA, Jain V, Wani AR, Rasool S, Nazir R, et al. Plant based natural products as potential ecofriendly and safer biopesticides: A comprehensive overview of their advantages over conventional pesticides, limitations and regulatory aspects. Microbial Pathogenesis. 2022;**173**:105854. DOI: 10.1016/j.micpath.2022.105854

[167] Werdin González JO, Gutiérrez MM, Ferrero AA, Fernández Band B. Essential oils nanoformulations for stored-product pest control – Characterization and biological properties. Chemosphere. 2014;**100**:130-138. DOI: 10.1016/j.chemosphere.2013.11.056

[168] Grumezescu AM, Holban AM, editors. Nutrients in Beverages. Amsterdam, Netherlands: Elsevier; 2019. ISBN 9780128168424

[169] Louro D, Fernandes JE, Quinot A. Current situation of tomato yellow leaf curl virus in Portugal. EPPO Bulletin. 2002;**32**:47-47. DOI: 10.1046/j.1365-2338.2002.00556_1.x

[170] Biowael A, Al Attar A, Othman B, Eldougdoug K. Frequency incidence of tomato chlorosis virus and tomato yellow leaf curl virus affecting tomato plants. Arab Universities Journal of Agricultural Sciences. 2024;**32**:25-40. DOI: 10.21608/ajs.2024.221291.1528

[171] Hasnain MS, Nayak AK. Natural Polysaccharides in Drug Delivery and Biomedical Applications. Amsterdam, Netherlands: Elsevier; 2019. ISBN 9780128170557

[172] Petitjean M, Isasi JR. Locust bean gum, a vegetable hydrocolloid with industrial and biopharmaceutical applications. Molecules. 2022;**27**:8265. DOI: 10.3390/molecules27238265

[173] Zhu B-J, Zayed MZ, Zhu H-X, Zhao J, Li S-P. Functional polysaccharides of carob fruit: A review. Chinese Medicine. 2019;**14**:40. DOI: 10.1186/s13020-019-0261-x

[174] Dionísio M, Grenha A. Locust bean gum: Exploring its potential for biopharmaceutical applications. Journal of Pharmacy & Bioallied Sciences. 2012;**4**:175. DOI: 10.4103/0975-7406.99013

[175] Preedy VR, Watson RR, Patel VB. Nuts and Seeds in Health and Disease Prevention. Amsterdam, Netherlands: Elsevier; 2011. ISBN 9780123756886

[176] Rodríguez-Solana R, Romano A, Moreno-Rojas JM. Carob pulp: A nutritional and functional by-product worldwide spread in the formulation of different food products and beverages. A review. PRO. 2021;**9**:1146. DOI: 10.3390/pr9071146

[177] Correia PJ, Pestana M. Sugars and phenols in carob tree fruits from different producing countries: A short review. Heliyon. 2024;**10**:e30922. DOI: 10.1016/j.heliyon.2024.e30922

[178] Ulukanli Z, Karabörklü S, Bozok F, Ates B, Erdogan S, Cenet M, et al. Chemical composition, antimicrobial, insecticidal, phytotoxic and antioxidant activities of Mediterranean *Pinus brutia* and *Pinus pinea* resin essential oils. Chinese Journal of Natural Medicines. 2014;**12**:901-910. DOI: 10.1016/S1875-5364(14)60133-3

[179] Moreira D, Gullón B, Gullón P, Gomes A, Tavaria F. Bioactive packaging using antioxidant extracts for the prevention of microbial food-spoilage. Food & Function. 2016;7:3273-3282. DOI: 10.1039/C6FO00553E

[180] Takwa S, Caleja C, Barreira JCM, Soković M, Achour L, Barros L, et al. *Arbutus unedo* L. and *Ocimum basilicum* L. as sources of natural preservatives for food industry: A case study using loaf bread. LWT. 2018;**88**:47-55. DOI: 10.1016/j.lwt.2017.09.041

[181] Serag MS et al. Essential oils as antimicrobial and food preservatives. In: Essential Oils - Advances in Extractions and Biological Applications. IntechOpen; 2022. DOI: 10.5772/intechopen.103000

[182] Bunse M, Daniels R, Gründemann C, Heilmann J, Kammerer DR, Keusgen M, et al. Essential oils as multicomponent mixtures and their potential for human health and well-being. Frontiers in Pharmacology. 2022;**13**:956541. DOI: 10.3389/fphar.2022.956541

[183] Ruas A, Graça A, Marto J, Gonçalves L, Oliveira A, da Silva AN, et al. Chemical characterization and bioactivity of commercial essential oils and hydrolates obtained from Portuguese forest logging and thinning. Molecules. 2022;**27**:3572. DOI: 10.3390/molecules27113572

[184] Ancuceanu R, Anghel AI, Hovaneţ MV, Ciobanu A-M, Lascu BE, Dinu M. Antioxidant activity of essential oils from *Pinaceae* species. Antioxidants. 2024;**13**:286. DOI: 10.3390/antiox13030286

[185] Ramos PAB, Pereira C, Gomes AP, Neto RT, Almeida A, Santos SAO, et al. Chemical characterisation, antioxidant and antibacterial activities of *Pinus pinaster* Ait. and *Pinus pinea* L. bark polar extracts: Prospecting forestry by-products as renewable sources of bioactive compounds. Applied Sciences. 2022;**12**:784. DOI: 10.3390/app12020784

[186] Alonso-Esteban JI, Carocho M, Barros D, Velho MV, Heleno S, Barros L. Chemical composition and industrial applications of maritime pine (*Pinus pinaster* Ait.) bark and other non-wood parts. Reviews in Environmental Science and Bio/Technology. 2022;**21**:583-633. DOI: 10.1007/s11157-022-09624-1

[187] Dorna H, Szopińska D, Rosińska A, Górski R. Chemical composition of fir, pine and thyme essential oils and their effect on onion (*Allium cepa* L.) seed quality. Agronomy. 2021;**11**:2445. DOI: 10.3390/agronomy11122445

[188] Faria JMS, Rodrigues AM. Metabolomic variability in the volatile composition of essential oils from Pinus Pinea and P. Pinaster. In: Proceedings of the the 1st International Electronic Conference on Biological Diversity, Ecology and Evolution. Basel Switzerland: MDPI; 2021. p. 14

[189] Silva N, Alves S, Gonçalves A, Amaral JS, Poeta P. Antimicrobial activity of essential oils from Mediterranean aromatic plants against several foodborne and spoilage Bacteria. Food Science and Technology International. 2013;**19**:503-510. DOI: 10.1177/1082013212442198

[190] Machado AM, Lopes V, Barata AM, Póvoa O, Farinha N, Figueiredo AC. Chemical variability of the essential oils from two Portuguese *Apiaceae*: *Coriandrum sativum* L. and *Foeniculum vulgare* mill. Plants. 2023;**12**:2749. DOI: 10.3390/plants12142749

[191] Zuzarte MR, Dinis AM, Canhoto J, Salgueiro L. Leaf trichomes of Portuguese *Lavandula* species: A comparative morphological study. Microscopy and Microanalysis. 2009;**15**:37-38. DOI: 10.1017/S143192760999064X

[192] Ordoudi SA, Papapostolou M, Nenadis N, Mantzouridou FT, Tsimidou MZ. Bay Laurel (*Laurus nobilis* L.) essential oil as a food preservative source: Chemistry, quality control, activity assessment, and applications to olive industry products. Food. 2022;**11**:752. DOI: 10.3390/foods11050752

[193] Preedy VR. Essential Oils in Food Preservation, Flavor and Safety. Amsterdam, Netherlands: Elsevier; 2016. ISBN 9780124166417

[194] Marongiu B, Piras A, Porcedda S, Falconieri D, Frau MA, Maxia A, et al. Antifungal activity and chemical composition of essential oils from *Smyrnium olusatrum* L. (*Apiaceae*) from Italy and Portugal. Natural Product Research. 2012;**26**:993-1003. DOI: 10.1080/14786419.2010.541876

[195] Gutiérrez TJ, editor. Polymers for Agri-Food Applications. Cham: Springer International Publishing; 2019. ISBN 978-3-030-19415-4

[196] da Costa Lopes AM, Silvestre AJD, Coutinho JAP. On the path to improve lignin depolymerization and functionalization into bio-based platform chemicals: A short review. Current Opinion in Green and Sustainable Chemistry. 2023;**43**:100850. DOI: 10.1016/j.cogsc.2023.100850

[197] Tavakolian M, Jafari SM, van de Ven TGM. A review on surface-functionalized cellulosic nanostructures as biocompatible antibacterial materials. Nano-Micro Letters. 2020;**12**:73. DOI: 10.1007/s40820-020-0408-4

[198] Farhat W, Venditti RA, Hubbe M, Taha M, Becquart F, Ayoub A. A review of water-resistant hemicellulose-based materials: Processing and applications. ChemSusChem. 2017;**10**:305-323. DOI: 10.1002/cssc.201601047

[199] Zhen L, Lange H, Zongo L, Crestini C. Chemical derivatization of commercially available condensed and hydrolyzable tannins. ACS Sustainable Chemistry & Engineering. 2021;**9**:10154-10166. DOI: 10.1021/acssuschemeng.1c02114

[200] Lin X, Shen T, Li M, Shaoyu J, Zhuang W, Li M, et al. Synthesis, characterization, and utilization of poly-amino acid-functionalized lignin for efficient and selective removal of lead ion from aqueous solution. Journal of Cleaner Production. 2022;**347**:131219. DOI: 10.1016/j.jclepro.2022.131219

[201] Protopopov AV, Kon′shin VV, Efryushin DD, Klevtsova MV. Chemical modification of lignins by amino acids. Chemistry of Natural Compounds. 2015;**51**:934-936. DOI: 10.1007/s10600-015-1451-0

[202] Amirjalayer S, Fuchs H, Marx D. Understanding the mechanocatalytic conversion of biomass: A low-energy one-step reaction mechanism by applying mechanical force. Angewandte Chemie International Edition. 2019;**58**:5232-5235. DOI: 10.1002/anie.201811091

[203] Do J-L, Friščić T. Mechanochemistry: A force of synthesis. ACS Central Science. 2017;**3**:13-19. DOI: 10.1021/acscentsci.6b00277

[204] Liu L-Y, Hua Q, Renneckar S. A simple route to synthesize esterified lignin derivatives. Green Chemistry. 2019;**21**:3682-3692. DOI: 10.1039/C9GC00844F

[205] Lê HQ, Zaitseva A, Pokki J, Ståhl M, Alopaeus V, Sixta H. Solubility of organosolv lignin in Γ-valerolactone/water binary mixtures. ChemSusChem. 2016;**9**:2939-2947. DOI: 10.1002/cssc.201600655

[206] Melro E, Filipe A, Valente AJM, Antunes FE, Romano A, Norgren M, et al. Levulinic acid: A novel sustainable solvent for lignin dissolution. International Journal of Biological Macromolecules. 2020;**164**:3454-3461. DOI: 10.1016/j.ijbiomac.2020.08.128

[207] Magalhães S, Filipe A, Melro E, Fernandes C, Vitorino C, Alves L, et al. Lignin extraction from waste pine sawdust using a biomass derived binary solvent system. Polymers (Basel). 2021;**13**:1090. DOI: 10.3390/polym13071090

[208] Li D, Qi L, Yang M, Gu Y, Xue Y, Chen J, et al. Switchable deep eutectic solvents for lignin dissolution and regeneration. Polymers (Basel). 2023;**15**:4233. DOI: 10.3390/polym15214233

[209] Fernandes C, Melro E, Magalhães S, Alves L, Craveiro R, Filipe A, et al. New deep eutectic solvent assisted extraction of highly pure lignin from maritime pine sawdust (*Pinus pinaster* Ait.). International Journal of Biological Macromolecules. 2021;**177**:294-305. DOI: 10.1016/j.ijbiomac.2021.02.088

[210] Leão S, Magalhães S, Alves L, Gamelas JAF, Lima C, Stein B, et al. Anionic bio-flocculants from sugarcane for purification of sucrose: An application of circular bioeconomy. Heliyon. 2023;**9**:e17134. DOI: 10.1016/j.heliyon.2023.e17134

[211] Grenda K, Gamelas JAF, Arnold J, Cayre OJ, Rasteiro MG. Cationization of *Eucalyptus* wood waste pulps with diverse lignin contents for potential application in colored wastewater treatment. RSC Advances. 2019;**9**:34814-34826. DOI: 10.1039/C9RA05757A

[212] Saito T, Kimura S, Nishiyama Y, Isogai A. Cellulose nanofibers prepared by TEMPO-mediated oxidation of native cellulose. Biomacromolecules. 2007;**8**:2485-2491. DOI: 10.1021/bm0703970

[213] Liimatainen H, Sirviö J, Pajari H, Hormi O, Niinimäki J. Regeneration and recycling of aqueous periodate solution

in dialdehyde cellulose production. Journal of Wood Chemistry and Technology. 2013;**33**:258-266. DOI: 10.1080/02773813.2013.783076

[214] Grenda K, Gamelas JAF, Arnold J, Cayre OJ, Rasteiro MG. Evaluation of anionic and cationic pulp-based flocculants with diverse lignin contents for application in effluent treatment From the textile industry: Flocculation monitoring. Frontiers in Chemistry. 2020;**8**:5. DOI: 10.3389/fchem.2020.00005

[215] Pedrosa JFS, Alves L, Neto CP, Rasteiro MG, Ferreira PJT. Assessment of the performance of cationic cellulose derivatives as calcium carbonate flocculant for papermaking. Polymers (Basel). 2022;**14**:3309. DOI: 10.3390/polym14163309

[216] Zhou Y, Saito T, Bergström L, Isogai A. Acid-free preparation of cellulose nanocrystals by TEMPO oxidation and subsequent cavitation. Biomacromolecules. 2018;**19**:633-639. DOI: 10.1021/acs.biomac.7b01730

[217] Wang Z-M, Li L, Xiao K-J, Wu J-Y. Homogeneous sulfation of bagasse cellulose in an ionic liquid and anticoagulation activity. Bioresource Technology. 2009;**100**:1687-1690. DOI: 10.1016/j.biortech.2008.09.002

[218] Watabe Y, Suzuki Y, Koike S, Shimamoto S, Kobayashi Y. Cellulose acetate, a new candidate feed supplement for ruminant animals: *In vitro* evaluations. Journal of Dairy Science. 2018;**101**:10929-10938. DOI: 10.3168/jds.2018-14969

[219] Koshani R, Tavakolian M, van de Ven TGM. Cellulose-based dispersants and flocculants. Journal of Materials Chemistry B. 2020;**8**:10502-10526. DOI: 10.1039/D0TB02021D

[220] Ribau Teixeira M, Ismail A, Medronho B, Alves L, Pedrosa JFS, Ferreira PJT, et al. Nanofibrillated cationic cellulose derivatives as flocculants for domestic wastewater treatment. Journal of Water Process Engineering. 2024;**58**:104817. DOI: 10.1016/j.jwpe.2024.104817

[221] Braz L, Grenha A, Corvo MC, Lourenço JP, Ferreira D, Sarmento B, et al. Synthesis and characterization of locust bean gum derivatives and their application in the production of nanoparticles. Carbohydrate Polymers. 2018;**181**:974-985. DOI: 10.1016/j.carbpol.2017.11.052

[222] Du X, Li J, Lindström ME. Modification of industrial softwood Kraft lignin using Mannich reaction with and without phenolation pretreatment. Industrial Crops and Products. 2014;**52**:729-735. DOI: 10.1016/j.indcrop.2013.11.035

[223] Ramadhoni B, Rifathin A, Yuliati F. Phosphorylation of alkaline lignin in methanesulfonic acid solvent and the use thereof as flame retardant of epoxy. Journal of The Institution of Engineers (India): Series D. 2023;**105**:1695-1702. DOI: 10.1007/s40033-023-00564-w

[224] Seddiqi H, Oliaei E, Honarkar H, Jin J, Geonzon LC, Bacabac RG, et al. Cellulose and its derivatives: Towards biomedical applications. Cellulose. 2021;**28**:1893-1931. DOI: 10.1007/s10570-020-03674-w

[225] Hasanin MS. Cellulose-based biomaterials: Chemistry and biomedical applications. Starch. 2022;**74**:2200060. DOI: 10.1002/star.202200060

[226] Pandey A. Pharmaceutical and biomedical applications of cellulose nanofibers: A review. Environmental

Chemistry Letters. 2021;**19**:2043-2055. DOI: 10.1007/s10311-021-01182-2

[227] Tudoroiu E-E, Dinu-Pîrvu C-E, Albu Kaya MG, Popa L, Anuța V, Prisada RM, et al. An overview of cellulose derivatives-based dressings for wound-healing management. Pharmaceuticals. 2021;**14**:1215. DOI: 10.3390/ph14121215

[228] Horue M, Silva JM, Berti IR, Brandão LR, da Barud HS, Castro GR. Bacterial cellulose-based materials as dressings for wound healing. Pharmaceutics. 2023;**15**:424. DOI: 10.3390/pharmaceutics15020424

[229] Sharma C, Bhardwaj NK. Bacterial nanocellulose: Present status, biomedical applications and future perspectives. Materials Science and Engineering: C. 2019;**104**:109963. DOI: 10.1016/j.msec.2019.109963

[230] Bianchi S, Bartoli F, Bruni C, Fernandez-Avila C, Rodriguez-Turienzo L, Mellado-Carretero J, et al. Opportunities and limitations in recycling fossil polymers from textiles. Macromolecules. 2023;**3**:120-148. DOI: 10.3390/macromol3020009

[231] Pakulska MM, Vulic K, Tam RY, Shoichet MS. Hybrid crosslinked methylcellulose hydrogel: A predictable and tunable platform for local drug delivery. Advanced Materials. 2015;**27**:5002-5008. DOI: 10.1002/adma.201502767

[232] Niemczyk-Soczynska B, Gradys A, Kolbuk D, Krzton-Maziopa A, Sajkiewicz P. Crosslinking kinetics of methylcellulose aqueous solution and its potential as a scaffold for tissue engineering. Polymers (Basel). 2019;**11**:1772. DOI: 10.3390/polym11111772

[233] Pasqui D, Torricelli P, De Cagna M, Fini M, Barbucci R. Carboxymethyl cellulose—hydroxyapatite hybrid hydrogel as a composite material for bone tissue engineering applications. Journal of Biomedial Materials Research Part A. 2014;**102**:1568-1579. DOI: 10.1002/jbm.a.34810

[234] Wang J, Liu W, Luo G, Li Z, Zhao C, Zhang H, et al. Synergistic effect of well-defined dual sites boosting the oxygen reduction reaction. Energy & Environmental Science. 2018;**11**:3375-3379. DOI: 10.1039/C8EE02656D

[235] He X, Lu W, Sun C, Khalesi H, Mata A, Andaleeb R, et al. Cellulose and cellulose derivatives: Different colloidal states and food-related applications. Carbohydrate Polymers. 2021;**255**:117334. DOI: 10.1016/j.carbpol.2020.117334

[236] Aziz T, Farid A, Haq F, Kiran M, Ullah A, Zhang K, et al. A review on the modification of cellulose and its applications. Polymers (Basel). 2022;**14**:3206. DOI: 10.3390/polym14153206

[237] Abbas HM, Abd El-Gawad MAM, Kassem JM, Salama M. Application of fat replacers in dairy products: A review. Foods and Raw Material. 2024;**12**:319-333. DOI: 10.21603/2308-4057-2024-2-612

[238] Shokri J, Adibkia K. Application of Cellulose and Cellulose Derivatives in Pharmaceutical Industries. London, UK, Rijeka: IntechOpen; 2013; p. Ch. 3

[239] Martins D, Rocha C, Dourado F, Gama M. Bacterial cellulose-carboxymethyl cellulose (BC: CMC) dry formulation as stabilizer and texturizing agent for surfactant-free cosmetic formulations. Colloids and Surfaces A: Physicochemical and Engineering

Aspects. 2021;**617**:126380. DOI: 10.1016/j.colsurfa.2021.126380

[240] Teo SH, Chee CY, Fahmi MZ, Wibawa Sakti SC, Lee HV. Review of functional aspects of nanocellulose-based pickering emulsifier for non-toxic application and its colloid stabilization mechanism. Molecules. 2022;**27**:7170. DOI: 10.3390/molecules27217170

[241] Li Q, Wang S, Jin X, Huang C, Xiang Z. The application of polysaccharides and their derivatives in pigment, barrier, and functional paper coatings. Polymers (Basel). 2020;**12**:1837. DOI: 10.3390/polym12081837

[242] Li M, Wu Q, Moon RJ, Hubbe MA, Bortner MJ. Rheological aspects of cellulose nanomaterials: Governing factors and emerging applications. Advanced Materials. 2021;**33**:2006052-2006080. DOI: 10.1002/adma.202006052

[243] Eid BM, Ibrahim NA. Recent developments in sustainable finishing of cellulosic textiles employing biotechnology. Journal of Cleaner Production. 2021;**284**:124701. DOI: 10.1016/j.jclepro.2020.124701

[244] Garba ZN, Lawan I, Zhou W, Zhang M, Wang L, Yuan Z. Microcrystalline cellulose (MCC) based materials as emerging adsorbents for the removal of dyes and heavy metals – A review. Science of the Total Environment. 2020;**717**:135070. DOI: 10.1016/j.scitotenv.2019.135070

[245] Akter M, Bhattacharjee M, Dhar AK, Rahman FBA, Haque S, Rashid TU, et al. Cellulose-based hydrogels for wastewater treatment: A concise review. Gels. 2021;**7**:30. DOI: 10.3390/gels7010030

[246] Ji SH, Yun JS. Natural cellulose-based multifunctional nanofibers for the effective removal of particulate matter and volatile organic compounds. Nanomaterials. 2023;**13**:1720. DOI: 10.3390/nano13111720

[247] Matei E, Predescu AM, Râpă M, Țurcanu AA, Mateș I, Constantin N, et al. Natural polymers and their nanocomposites used for environmental applications. Nanomaterials. 2022;**12**:1707. DOI: 10.3390/nano12101707

[248] Zainul Armir NA, Zulkifli A, Gunaseelan S, Palanivelu SD, Salleh KM, Che Othman MH, et al. Regenerated cellulose products for agricultural and their potential: A review. Polymers (Basel). 2021;**13**:3586. DOI: 10.3390/polym13203586

[249] Saberi Riseh R. Advancing agriculture through bioresource technology: The role of cellulose-based biodegradable mulches. International Journal of Biological Macromolecules. 2024;**255**:128006. DOI: 10.1016/j.ijbiomac.2023.128006

[250] Jin K, Tang Y, Liu J, Wang J, Ye C. Nanofibrillated cellulose as coating agent for food packaging paper. International Journal of Biological Macromolecules. 2021;**168**:331-338. DOI: 10.1016/j.ijbiomac.2020.12.066

[251] Bhardwaj A, Alam T, Sharma V, Alam MS, Hamid H, Deshwal GK. Lignocellulosic agricultural biomass as a biodegradable and eco-friendly alternative for polymer-based food packaging. Journal of Packaging Technology and Reseach. 2020;**4**:205-216. DOI: 10.1007/s41783-020-00089-7

[252] Gupta P, Toksha B, Rahaman M. A review on biodegradable packaging films from vegetative and food waste. Chemical Record. 2022;**22**:e202100326. DOI: 10.1002/tcr.202100326

[253] Dahlström C, Duan R, Eivazi A, Magalhães S, Alves L, Engholm M, et al. Stacking self-gluing cellulose II films: A facile strategy for the formation of novel all-cellulose laminates. Carbohydrate Polymers. 2024;**344**:122523. DOI: 10.1016/j.carbpol.2024.122523

[254] Jiang X, Bai Y, Chen X, Liu W. A review on raw materials, commercial production and properties of lyocell fiber. Journal of Bioresources and Bioproducts. 2020;**5**:16-25. DOI: 10.1016/j.jobab.2020.03.002

[255] Vicente TA, Araújo A, Mendes MJ, Nunes D, Oliveira MJ, Sanchez-Sobrado O, et al. Multifunctional cellulose-paper for light harvesting and smart sensing applications. Journal of Materials Chemistry C. 2018;**6**:3143-3181. DOI: 10.1039/C7TC05271E

[256] Zhao D, Zhu Y, Cheng W, Chen W, Wu Y, Yu H. Cellulose-based flexible functional materials for emerging intelligent electronics. Advanced Materials. 2021;**33**:2000619. DOI: 10.1002/adma.202000619

[257] Miyashiro D, Hamano R, Umemura K. A review of applications using mixed materials of cellulose, nanocellulose and carbon nanotubes. Nanomaterials. 2020;**10**:186. DOI: 10.3390/nano10020186

[258] Thiemann S, Sachnov SJ, Pettersson F, Bollström R, Österbacka R, Wasserscheid P, et al. Cellulose-based ionogels for paper electronics. Advanced Functional Materials. 2014;**24**:625-634. DOI: 10.1002/adfm.201302026

[259] Wasim M, Shi F, Liu J, Khan MR, Farooq A, Sanbhal N, et al. Extraction of cellulose to progress in cellulosic nanocomposites for their potential applications in supercapacitors and energy storage devices. Journal of Materials Science. 2021;**56**:14448-14486. DOI: 10.1007/s10853-021-06215-3

[260] Dahlström C, Eivazi A, Nejström M, Zhang R, Pettersson T, Iftikhar H, et al. Regenerated cellulose properties tailored for optimized triboelectric output and the effect of counter-tribolayers. Cellulose. 2024;**31**:2047-2061. DOI: 10.1007/s10570-024-05745-8

[261] From M, Larsson PT, Andreasson B, Medronho B, Svanedal I, Edlund H, et al. Tuning the properties of regenerated cellulose: Effects of polarity and water solubility of the coagulation medium. Carbohydrate Polymers. 2020;**236**:116068. DOI: 10.1016/j.carbpol.2020.116068

[262] Zhang R, Dahlström C, Zou H, Jonzon J, Hummelgård M, Örtegren J, et al. Cellulose-based fully green triboelectric nanogenerators with output power density of 300 W m −2. Advanced Materials. 2020;**32**:2002824. DOI: 10.1002/adma.202002824

[263] Feldman D. Lignin nanocomposites. Journal of Macromolecular Science, Part A. 2016;**53**:382-387. DOI: 10.1080/10601325.2016.1166006

[264] Luo H, Abu-Omar MM. Chemicals From lignin. In: Encyclopedia of Sustainable Technologies. Amsterdam, Netherlands: Elsevier; 2017. pp. 573-585

[265] Reesi F, Minaiyan M, Taheri A. A novel lignin-based nanofibrous dressing containing arginine for wound-healing applications. Drug Delivery and Translational Research. 2018;**8**:111-122. DOI: 10.1007/s13346-017-0441-0

[266] Zhang Y, Cheng G. Lignin-containing hydrogels: Transforming a low-value byproduct to absorbents, wound dressings and strain sensors.

Chemical Engineering Journal. 2023;**475**:146460. DOI: 10.1016/j.cej.2023.146460

[267] Dai M, Zhang J, Liu N, Zhang X-H. Precise engineering of lignin incorporated dextran/glycol nanomaterials for wound dressings for the care of anorectal surgery. Journal of Polymers and the Environment. 2022;**30**:206-216. DOI: 10.1007/s10924-021-02175-6

[268] Fernandes C, Medronho B, Alves L, Rasteiro MG. On hair care physicochemistry: From structure and degradation to novel biobased conditioning agents. Polymers (Basel). 2023;**15**:608. DOI: 10.3390/polym15030608

[269] Morganti P, Morganti G. Natural polymers for natural hair: The smart use of an innovative nanocarrier. In: Nanocosmetics. Amsterdam, Netherlands: Elsevier; 2020. pp. 267-285

[270] Morganti P, Coltelli M-B. A new carrier for advanced cosmeceuticals. Cosmetics. 2019;**6**:10. DOI: 10.3390/cosmetics6010010

[271] Garg J, Nee Chiu M, Krishnan S, Kumar Tripathi L, Pandit S, Farasati Far B, et al. Applications of lignin nanoparticles for cancer drug delivery: An update. Materials Letters. 2022;**311**:131573. DOI: 10.1016/j.matlet.2021.131573

[272] Figueiredo P, Lintinen K, Kiriazis A, Hynninen V, Liu Z, Bauleth-Ramos T, et al. *In vitro* evaluation of biodegradable lignin-based nanoparticles for drug delivery and enhanced antiproliferation effect in cancer cells. Biomaterials. 2017;**121**:97-108. DOI: 10.1016/j.biomaterials.2016.12.034

[273] Van DN, Tsygankov PY, Menshutina NV. Facile lignin extraction and application as natural UV blockers in cosmetic formulations. ChemEngineering. 2024;**8**:69. DOI: 10.3390/chemengineering8040069

[274] Silva MNB, Scopel E, Rezende CA. From black liquor to tinted sunscreens: Washing out Kraft lignin unpleasant odor and improving its properties by lignin nanoparticle preparation. Industrial Crops and Products. 2024;**218**:118910. DOI: 10.1016/j.indcrop.2024.118910

[275] Liu B, Zhang W, Zeng J, Gong N, Ying G, Li P, et al. Acid-catalyzed phenolation of lignin with tea polyphenol: Enhancing UV resistance and oxidation resistance for potential applications. International Journal of Biological Macromolecules. 2024;**267**:131462. DOI: 10.1016/j.ijbiomac.2024.131462

[276] Xia Z, You J, An H, Wang Y, Zhang J, Yin C, et al. Eco-friendly fractionation of natural straws: Sustainable ultralong room-temperature phosphorescence and super anti-ultraviolet materials. Science China. Chemistry. 2024;**67**:2373-2381. DOI: 10.1007/s11426-024-2016-8

[277] Lyu Y, Ji X, Tian Z, Ji H, Zhang F, Zhou J. Extraction of glyoxylic acid stabilized lignin from lignocellulosic biomass for a natural sunscreen additive. ACS Sustainable Resource Management Journal. 2024;**1**:757-766. DOI: 10.1021/acssusresmgt.3c00138

[278] Fan Y, Ji H, Ji X, Tian Z, Chen J. Lignocellulosic biomass pretreatment with a lignin stabilization strategy and valorization toward multipurpose fractionation. International Journal of Biological Macromolecules. 2024;**259**:129186. DOI: 10.1016/j.ijbiomac.2023.129186

[279] Cheng J, Zhou X, Huang C, Yoo CG, Meng X, Fang G, et al.

Low-chromophore lignin isolation from natural biomass with polyol-based deep eutectic solvents. Green Chemistry. 2024;**26**:8298-8314. DOI: 10.1039/D4GC01824A

[280] Girard V, Fragnières L, Chapuis H, Brosse N, Marchal-Heussler L, Canilho N, et al. The impact of lignin biopolymer sources, isolation, and size reduction from the macro- to nanoscale on the performances of next-generation sunscreen. Polymers (Basel). 2024;**16**:1901. DOI: 10.3390/polym16131901

[281] Mukheja Y, Kaur J, Pathania K, Sah SP, Salunke DB, Sangamwar AT, et al. Recent advances in pharmaceutical and biotechnological applications of lignin-based materials. International Journal of Biological Macromolecules. 2023;**241**:124601. DOI: 10.1016/j.ijbiomac.2023.124601

[282] Aadil KR, Mussatto SI, Jha H. Synthesis and characterization of silver nanoparticles loaded poly(vinyl alcohol)-lignin electrospun nanofibers and their antimicrobial activity. International Journal of Biological Macromolecules. 2018;**120**:763-767. DOI: 10.1016/j.ijbiomac.2018.08.109

[283] Lu F, Ni S, Meng X, Si C, Qin M, Wang Y, et al. Preparation of pH-sensitive lignin nanoparticles and its application in hydrophobic drug delivery. Industrial Crops and Products. 2023;**202**:117012. DOI: 10.1016/j.indcrop.2023.117012

[284] Barapatre A, Meena AS, Mekala S, Das A, Jha H. In vitro evaluation of antioxidant and cytotoxic activities of lignin fractions extracted from Acacia Nilotica. International Journal of Biological Macromolecules. 2016;**86**:443-453. DOI: 10.1016/j.ijbiomac.2016.01.109

[285] Figueiredo P, Ferro C, Kemell M, Liu Z, Kiriazis A, Lintinen K, et al. Functionalization of Carboxylated lignin nanoparticles for targeted and PH-responsive delivery of anticancer drugs. Nanomedicine. 2017;**12**:2581-2596. DOI: 10.2217/nnm-2017-0219

[286] Alqahtani MS, Alqahtani A, Al-Thabit A, Roni M, Syed R. Novel lignin nanoparticles for Oral drug delivery. Journal of Materials Chemistry B. 2019;**7**:4461-4473. DOI: 10.1039/C9TB00594C

[287] Li Y, Yang D, Wang S, Xu H, Li P. Fabrication and optimization of Pickering emulsion stabilized by lignin nanoparticles for curcumin encapsulation. ACS Omega. 2024;**9**:21994-22002. DOI: 10.1021/acsomega.3c10395

[288] Dai L, Liu R, Hu L-Q, Zou Z-F, Si C-L. Lignin nanoparticle as a novel Green carrier for the efficient delivery of resveratrol. ACS Sustainable Chemistry & Engineering. 2017;**5**:8241-8249. DOI: 10.1021/acssuschemeng.7b01903

[289] Dai L, Zhu W, Liu R, Si C. Lignin-containing self-nanoemulsifying drug delivery system for enhance stability and Oral absorption of Trans -resveratrol. Particle and Particle Systems Characterization. 2018;**35**:1700447. DOI: 10.1002/ppsc.201700447

[290] Rahdar A, Sargazi S, Barani M, Shahraki S, Sabir F, Aboudzadeh M. Lignin-stabilized doxorubicin microemulsions: Synthesis, physical characterization, and In vitro assessments. Polymers (Basel). 2021;**13**:641. DOI: 10.3390/polym13040641

[291] Zhou Y, Han Y, Li G, Yang S, Xiong F, Chu F. Preparation of targeted lignin–based hollow nanoparticles for

the delivery of doxorubicin. Nanomaterials. 2019;**9**:188. DOI: 10.3390/nano9020188

[292] Chaudhary M, Sinha VR. Lignin-based carriers for drug delivery applications: From an industrial waste to a pharmaceutical aid. European Polymer Journal. 2023;**195**:112206. DOI: 10.1016/j.eurpolymj.2023.112206

[293] Magalhães S, Alves L, Medronho B, Romano A, da Rasteiro MG. Microplastics in ecosystems: From current trends to bio-based removal strategies. Molecules. 2020;**25**:3954. DOI: 10.3390/molecules25173954

[294] Guo K, Gao B, Yue Q, Xu X, Li R, Shen X. Characterization and performance of a novel lignin-based Flocculant for the treatment of dye wastewater. International Biodeterioration & Biodegradation. 2018;**133**:99-107. DOI: 10.1016/j.ibiod.2018.06.015

[295] Chen J, Eraghi Kazzaz A, Alipoor Mazandarani N, Hosseinpour Feizi Z, Fatehi P. Production of flocculants, adsorbents, and dispersants from lignin. Molecules. 2018;**23**:868. DOI: 10.3390/molecules23040868

[296] Wang S, Kong F, Gao W, Fatehi P. Novel process for generating cationic lignin based Flocculant. Industrial and Engineering Chemistry Research. 2018;**57**:6595-6608. DOI: 10.1021/acs.iecr.7b05381

[297] Wang T, Li H, Diao X, Lu X, Ma D, Ji N. Lignin to dispersants, adsorbents, flocculants and adhesives: A critical review on industrial applications of lignin. Industrial Crops and Products. 2023;**199**:116715. DOI: 10.1016/j.indcrop.2023.116715

[298] Moore C, Gao W, Fatehi P. Cationic lignin polymers as Flocculant for municipal wastewater. Polymers (Basel). 2021;**13**:3871. DOI: 10.3390/polym13223871

[299] Wang H, Song J, Yan M, Li J, Yang J, Huang M, et al. Waste lignin-based cationic flocculants treating dyeing wastewate: rFabrication, performance, and mechanism. Science of the Total Environment. 2023;**874**:162383. DOI: 10.1016/j.scitotenv.2023.162383

[300] Wang B, Cao X-F, Yu S-X, Sun Z-H, Shen X-J, Wen J-L, et al. New sustainable mannich-functionalized lignin flocculants for ultra-efficiently tailoring wastewater purification. Journal of Cleaner Production. 2023;**387**:135801. DOI: 10.1016/j.jclepro.2022.135801

Chapter 9

General Aspects and Keypoints in Mathematical Modeling of Biomass Hydrothermal Pretreatment Processes

Samuel Oliveira, Fernando Paz-Cedeno, Naila Mori, Adilson Gonçalves and Fernando Masarin

Abstract

In this chapter, the use of biomass as a raw material for the biotechnological obtainment of various products within the scope of biorefineries was considered, highlighting the pretreatment of biomass as an essential step in the overall biomass conversion process, as well as presenting those most used for this purpose, including physical and chemical methods. Generalities about mathematical models of pretreatment processes, what they consist of, and their applications were also reported. A comprehensive description of hydrothermal pretreatment processes, their main characteristics, and variables was provided. Different approaches to mathematical modeling of hydrothermal processes were analyzed, including (i) modeling based on mechanisms and kinetics; (ii) modeling using the severity factor as a reaction coordinate that incorporates the main variables that affect the hydrothermal process; (iii) modeling through the application of statistical experimental designs and multiple linear regression to optimize the hydrothermal process. Finally, data from the literature and our research group regarding the application of the third modeling approach were presented and discussed.

Keywords: biorefinery, biomass hydrothermal pretreatment, mathematical models, kinetics, optimization

1. Introduction

Lignocellulosic biomass can be considered a raw material for the obtainment of bioproducts in the medium and long term, due to its low cost and high availability [1]. However, obtaining bioproducts from lignocellulosic materials is not as simple as those generated from materials rich in sugar or starch. Furthermore, the lignocellulosic conversion involves other steps, including biomass pretreatment, which is considered a crucial step.

Several studies have been carried out on this step for different types of lignocellulosic waste, such as rice and wheat straw, sunflower stalks, and sugarcane bagasse

[2–8]. However, the cost of some types of pretreatments and the low density of the lignocellulosic material still make it uneconomical to obtain certain products, especially those with lower added value.

Conceptually, a biorefinery is defined as the integration of biomass conversion processes and equipment to obtain various bioproducts and energy. From an eco-efficiency viewpoint, the biorefinery demonstrates better environmental performance, mainly in the potential for global warming, due to CO_2 fixation during acid fermentation [9].

Regardless of the use of the main fractions contained in the biomass, preliminary processing is necessary to separate them, especially lignin, which can be considered the most difficult physical barrier, represented by the fibers of these materials cemented together. Separation of the main macromolecular components from bagasse and straw, for example, could take part of this residue to the production of chemical compounds such as sugars, alcohols, organic acids, furfural, phenols, and other aromatic compounds, of greater economic value than that of raw biomass, which could be obtained through chemical, biochemical, physical, or combinations between them [10].

Therefore, pretreatment serves as the initial step in the biomass conversion process. Its primary objective is to modify and remove components that impede access to the substrate, thereby facilitating chemical or enzymatic reactions. As a consequence, there is an increase in the hydrolysis yield and in the amount of fermentable sugars and other intermediate substrates of interest. Studies have shown that pretreatment is the most significant step, and decisive factor in making the technology for obtaining products from biomass, this is because this step defines the yield and cost at which cellulose and hemicellulose can be converted into bioproducts [11]. Furthermore, the impact of pretreatment on the cost of the subsequent product recovery must also be considered [10, 12].

Many pretreatment methods have been studied or are still being developed. It is difficult to evaluate and compare pretreatment technologies due to various pre- and post-processing costs, investment capital, recovery of chemical reagents, and waste treatment systems [13]. However, as an integral part of an industrial system, the biorefinery, a mass balance analysis can be used to validate the efficiency of a process for a given raw material [12]. The methods of pretreatment depend on the material used and the purpose of the use of lignocellulosic fractions, including physical, chemical, physical-chemical, and biological methods.

Regarding the physical methods, the most used for bioproducts obtainment from agroindustrial residues are combinations of methods such as grinding or chipping, which reduce cellulose crystallinity and improve the efficiency of the downstream process [14]. In addition to these methods, other physical pretreatments have been widely studied such as hydrothermal, microwave irradiation, by ultrasound, pyrolysis, and extrusion. However, physical processes, in general, are very expensive and require more energy inputs, and, thinking about a process on a high scale (commercial production), would hardly be employed [13].

Regarding chemical and physical-chemical methods, there are several possibilities, but, the most studied are generally those that involve the use of reagents such as sodium hydroxide (to remove mostly lignin), perchloric, peracetic, sulfuric, phosphoric, and hydrochloric acids (mainly for hemicellulose removal), in addition to organic solvents [13, 15]. However, the greater resistance that exists regarding the use of these chemical reagents is the concern with the total cost of bioconversion of cellulosic biomass [13], showing that among chemical treatments, pretreatment with diluted sulfuric acid is the most studied and the most popular. However, there are other types that stand out, such as steam explosion treatment (catalyzed or not), AFEX (ammonia fiber expansion/explosion), etc. [16].

The biological pretreatment method consists of the use of microorganisms such as white, brown, and soft rot fungi for the decomposition of the cellulose complex, releasing the cellulose for the stage subsequent to biomass conversion [14, 16, 17]. The advantages of this pretreatment method are that it does not use chemical reagents and low energy consumption since there is no need for large mechanical supports [11]. As disadvantages are the low reaction rates and low yields, which prevents, in fact, the implementation [11].

Hydrothermal pretreatment of lignocellulosic materials for bioproduct obtainment has gained great prominence and importance in recent years [18]. Water under high pressure and temperature can penetrate the biomass, hydrate the cellulose, and solubilize the hemicellulose, in addition to removing part of the lignin. The primary advantages of this treatment are its lack of chemical reagents and the elimination of the need for corrosion-resistant materials in constructing the reactor used for hydrolysis. Another key advantage is that there is no need to reduce the size of the biomass beforehand. Additionally, this method requires only a small amount of chemicals for neutralizing the hydrolysate and produces less waste compared to other pretreatment methods [19, 20]. The main objective of hydrothermal pretreatment is to solubilize the hemicelluloses, making cellulose more accessible for enzymatic hydrolysis. Among the macromolecular components of lignocellulosic materials, the hemicelluloses have greater chemical and thermal sensitivity [21, 22] and therefore at high temperatures (e.g. 180°C) under neutral conditions, initially, solubilization of hemicelluloses occurs followed by small fractions of lignin [23, 24]. The principle of hydrothermal pretreatment is based on the autoionization of water caused by high temperatures. The formation of hydrogen ions from the autoionization of water promotes acidification of the medium and induces the solubilization and hydrolysis of the acetyl groups of hemicellulose. Acetic acid is a by-product of this reaction, which acts as a catalyst for hydrolysis [25]. During the exothermic reaction of solubilization of hemicelluloses, acids are formed, which self-catalyze the reaction of hydrolysis of hemicelluloses.

The economic impact of the pretreatment step on the entire biomass conversion process (which can be up to a third of the total processing cost) has stimulated the development of useful mathematical models for design and optimization purposes [26].

The most commonly developed models for the pretreatment processes are kinetic models, assuming a first-order dependence of reaction rates on biomass components and an Arrhenius-type relationship to describe the rate constants as a function of the temperature [27]. Mathematical models of pretreatment processes are conceived based on both experimental and literature data, aiming to identify the kinetic expressions involved in the different reactions and incorporate them into mass balances to quantify the consumed and generated amounts of the different participating components [26]. Kinetics-based models are classified as appropriate for optimization studies because they have a superior ability to evaluate and interpret the effects of independent variables on the pretreatment process [26].

The kinetic models must be as simple as possible to enable them to explain in a satisfactory way the overall pretreatment process [26]. However, owing to the heterogeneous nature of the reactions involved, various alternative modeling approaches for describing the behavior of the systems have been used, including severity factor, Design of Experiments (DOE) coupled with multivariate linear regression, artificial neural network, and fuzzy inference [27].

Agroindustrial residues (sugarcane bagasse, wheat straw, rice straw, corn stover, etc.) constitute the principal source of lignocellulosic biomass, which, in turn, is mainly composed of cellulose, hemicellulose, and lignin. Cellulose is the main source of fermentable sugars for ethanol production from biomass, being protected by a network of lignin and hemicellulose, which requires an adequate pretreatment to remove the protecting shield, making the cellulose more susceptible to enzymatic digestion [28]. Hemicellulose is a polysaccharide composed of different types of monosaccharides, including xylose, arabinose, galactose, and other compounds [28].

Autohydrolysis of biomass by a hydrothermal process with liquid hot water or high-pressure steam is a preliminary step in the obtainment of hemicellulose and its derivatives for the manufacture of materials and chemical products, being a very complex reaction system whose kinetic study is an essential requisite to design lignocellulose pretreatment reactors [28]. The distribution of products in autohydrolysis strongly depends on the operating conditions of temperature and reaction time, making it necessary to optimize these variables through mathematical models. Therefore, it is necessary to understand the autohydrolysis mechanism and the associated kinetics.

The objective of this chapter is to analyze some mathematical models of biomass hydrothermal pretreatment processes, focusing on their general aspects and key points, with a view to developing and optimizing technology for biomass biorefineries.

2. Biomass hydrothermal pretreatment processes

To gain a comprehensive understanding of the biomass hydrothermal pretreatment process, it is crucial to examine the chemical structure of hemicellulose in detail. Hemicellulose, a complex carbohydrate that forms a significant component of plant cell walls, plays a vital role in the overall composition and behavior of lignocellulosic biomass. By analyzing its unique chemical structure, it is possible to elucidate how hemicellulose interacts with other biopolymers, such as cellulose and lignin, during the pretreatment step. This knowledge can help optimize the hydrothermal process, ultimately improving the efficiency of biomass conversion into valuable bioproducts. In the next subsection, the chemical structure of hemicellulose in lignocellulosic biomass will be discussed, emphasizing its primary component, the xylan backbone.

2.1 Chemical structure of hemicellulose

The hemicellulose from hardwoods and grasses consists of homopolymers known as arabinoxylan, as it has a main chain composed of anhydroxylose units (**Figure 1**).

Arabinoxylan is entirely amorphous and contains side groups (arabinose, acetic acid, and methyl glucuronic acid) along its main chain (**Figure 1**). The degree of polymerization of arabinoxylan ranges between 100 and 200 units, and it is situated within a matrix of cellulose and lignin in the plant cell wall, forming a lignin-carbohydrate complex. Although arabinoxylan is more easily degradable than cellulose, it still provides significant resistance due to its complex structural network [29–31].

The main chemical bonds between lignin and arabinoxylan molecules in hardwoods are illustrated in **Figure 2**.

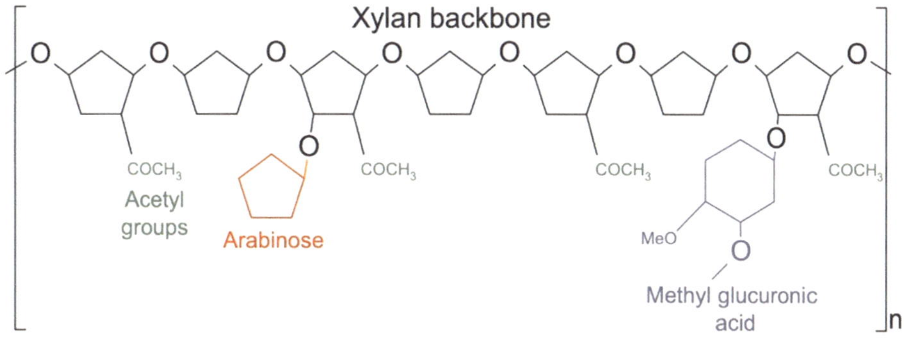

Figure 1.
Molecular structure of L-arabino-D-xylan, the main hemicellulose component of hardwoods and grasses (source: own authors).

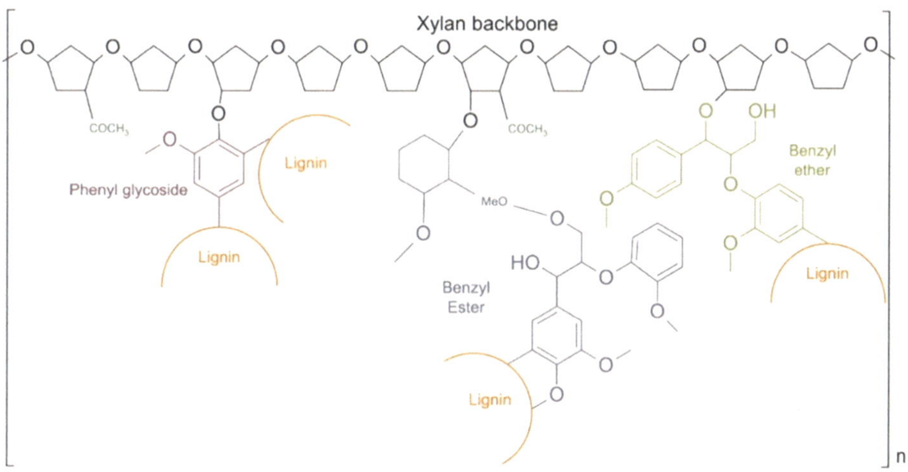

Figure 2.
Diagram illustrating phenolic glycosidic ester and benzyl ether and ester bonds residues attached to lignin (source: own authors).

The presence of these bonds has been confirmed in the lignin-carbohydrate complex isolated from pine wood (coniferous wood) using two-dimensional nuclear magnetic resonance (2D-NMR). **Figure 2** depicts an ester phenolic glycosidic, ester, and ether benzyl bonds to lignin, respectively. In the case of the phenolic glycosidic bond, the corresponding area in the spectrum was integrated, indicating that this bond is present at approximately 8 units per 100 C_9 units of lignin, making it the most frequent bond [29, 31, 32].

The primary chemical bonds between lignin and arabinoxylan molecules in grasses are illustrated in **Figure 3**.

In grasses, lignin is connected to xylan through ferulic and p-coumaric acids, which are phenolic acids linked at the C5 position of an arabinose molecule via an ester bond. Arabinose is coupled to xylan through an ether-type linkage (arabinose group). Additionally, lignin is connected to phenolic acids via propanoic chains at the α-carbon, primarily through ether-type bonds. The phenolic acids are attached to lignin through radical processes; however, ferulic and p-coumaric acids are coupled only to monomers (coniferol and sinapyl alcohols) and not lignin oligomers. The primary

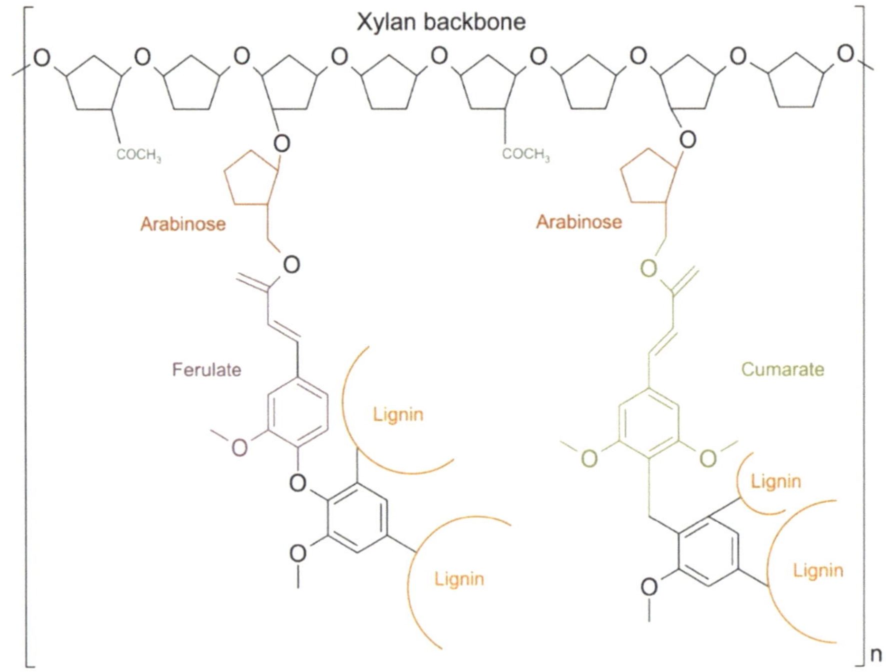

Figure 3.
Diagram illustrating ferulate residues attached to lignin (source: own authors).

function of these acids is to act as nucleation points during the biosynthesis of lignin in grasses [29, 33, 34]. Sugarcane bagasse contains approximately 1.7% ferulic acid and 7% p-coumaric acid (% dry weight) [35].

2.2 Hydrothermal pretreatment

Hydrothermal pretreatment is a technology in which lignocellulosic materials, particularly hardwoods and grasses, are treated with water at high temperatures (between 140 and 220°C) and pressures (ranging from 5 to 20 $kgf.cm^{-2}$), under a specific mass load or consistency (the ratio of water to the dry weight of lignocellulosic biomass), promoting the deconstruction of the lignocellulosic material (**Figure 4**).

Hydrothermal pretreatment is an efficient and environmentally sustainable technique for preparing lignocellulosic biomass to obtain various bioproducts. It stands out by not requiring additional chemicals, simplifying the process, and reducing operational costs, while also avoiding equipment corrosion [36]. However, this method requires high temperatures which can result in the formation of compounds that inhibit biotechnological processes [37].

The hydrothermal pretreatment process lowers the pH of the reaction medium due to the release of acetic acid from the side chains of arabinoxylan (**Figure 1**). Moreover, at higher temperatures (150–230°C), the hydrogen bonds in water weaken, resulting in its autoionization into hydronium ions (H_3O^+), which act as catalysts and further reduce the pH of the medium. This pH decrease promotes the hydrolysis and partial solubilization of the hemicellulose fraction, like what occurs in dilute acid

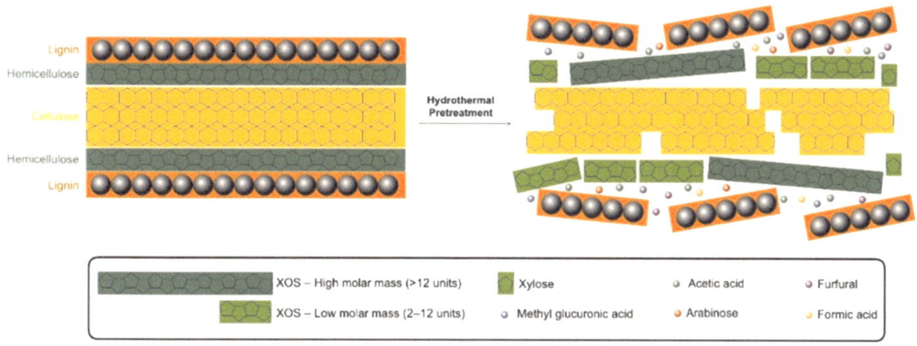

Figure 4.
Model of lignocellulosic material deconstruction after hydrothermal pretreatment application (source: own authors).

pretreatment, but to a lesser extent, thereby preserving the cellulose fraction and resulting in pulp recovery yields of 80–90% [36].

The main fraction of lignocellulosic materials (hardwoods and grasses) hydrolyzed during hydrothermal pretreatment is arabinoxylan because it is the most susceptible of the three major components of the cell wall (**Figures 1–3**). This pretreatment is more selective in the hydrolysis of this polysaccharide [31, 38–42]. Hemicellulose degradation occurs in three main stages. Initially, reactions occur on the surface of the material, resulting in the formation of primary products (high molecular weight oligosaccharides). These are subsequently broken down into xylooligosaccharides (XOS) (low molecular weight oligosaccharides) and, eventually, xylose (**Figure 4**). Additionally, methyl glucuronic and acetic acids are released due to the cleavage of ether and ester bonds, respectively (**Figure 1**). Under more severe conditions of temperature and time, XOS can be hydrolyzed to xylose, which, in acidic environments and at high temperatures, can be dehydrated to furfural and further oxidized to formic acid [37, 43, 44]. The efficiency of pretreatment can be enhanced by optimizing reaction conditions (temperature, time, and consistency), aiming to minimize the formation of inhibitory compounds for biotechnological processes and/or maximize the formation of XOS [36, 43, 44], which have prebiotic potential and various industrial applications.

Hydrothermal pretreatment is a technically viable option for producing XOS (soluble hemicellulose fraction). Additionally, within the biorefinery concept, the recovered pulp (insoluble fraction rich in cellulose and lignin) after hydrothermal pretreatment can be enzymatically hydrolyzed by cellulases, resulting in a cellulosic hydrolysate (soluble fraction), i.e., can be converted into bioethanol. This is particularly effective when combined with other techniques, such as ensiling, which can increase sugar recovery and improve ethanol yields by up to 10%. Lastly, the final residue recovered from the enzymatic hydrolysis step (insoluble fraction) can be refined to obtain technical lignin, which can be used for the formulation of chemicals, adhesives, and carbon fibers, among other applications [36, 43, 44].

In studies conducted by our research group, the optimized conditions for time, temperature, and mass load or consistency for hydrothermal pretreatment to produce XOS from sugarcane bagasse, straw, and wood by-products were 75.3 min at 161.98°C with 10% (w/v) consistency, 64.9 min at 177.25°C with 10% (w/v) consistency, and 65 min at 161°C with 10% (w/v) consistency, respectively [43, 44]. Under these optimized conditions, XOS yields were 74.9, 59.5, and 42.6%, respectively, resulting in the production of 53.3, 96.3, and 60.0 mg of XOS per gram of dry by-product. These results were confirmed through mass balance data, considering that the remaining pulp mass represents the hemicellulosic fraction not converted into bioproducts. This unconverted fraction was added to the components quantified in the hemicellulosic hydrolysate, completing the recovered mass, with the observed differences representing unquantified components.

3. Mathematical modeling of biomass hydrothermal pretreatment processes

In this section, some approaches commonly employed for mathematical modeling of biomass hydrothermal pretreatment processes are presented.

3.1 Mathematical modeling based on mechanisms and kinetics

The autohydrolysis of hemicelluloses has aroused great interest, and kinetic models based on different reaction schemes can be found in the literature [45]. In addition to raw materials, the main variables investigated in autohydrolysis processes are temperature, time, particle size, solid-to-liquid ratio, and pH [46].

A common modeling approach is to use formal kinetics to represent system behavior according to reaction mechanisms established from experimental data [45]. Generally, a sequence of consecutive hydrolytic reactions is considered, including [45]: hydrolysis of hemicellulose in the solid biomass to soluble oligomers, which are subsequently depolymerized in the liquid phase to monosaccharides, including those of six-carbon, which, in turn, are degraded to hydroxymethylfurfural (HMF). Xylose and arabinose are also dehydrated to furfural. Finally, furfural and HMF are condensed and degraded into other products. Such reactions are initiated by the hydronium ion resulting from the autoionization of water and are further accelerated by the generation of free uronic acids from the hydrolysis of the uronic and acetyl groups in hemicellulose [45].

Although most of the proposed mathematical models are based on the classical Saeman's model, which was developed for dilute acid hydrolysis of cellulose and hemicellulose, this model was used to describe the pretreatment of sugarcane bagasse with liquid hot water, considering the following pseudo-homogeneous, series-parallel, irreversible, and first-order reactions [26]:

$$Cellulose \overset{k_1'}{\rightarrow} Glucose \overset{k_2'}{\rightarrow} Degradation\ Products\ (\mathrm{DP}) \tag{1}$$

$$Hemicellulose \overset{k_1''}{\rightarrow} Monomers \overset{k_2''}{\rightarrow} Degradation\ Products\ (\mathrm{DP}) \tag{2}$$

Saeman's model can be applied to other polysaccharides, as follows:

$$Polymer\ (\mathrm{P}) \overset{k_1}{\rightarrow} Monomer\ (\mathrm{M}) \overset{k_2}{\rightarrow} Degradation\ Products\ (\mathrm{DP}) \tag{3}$$

According to Eqs. (1)-(3), the polymers (P) (e.g., glucan, xylan, araban, or galactan) contained in the initial biomass are converted into monomers (M) (e.g. glucose, xylose, arabinose, or galactose) which, in turn, are subsequently converted

Mass balance	Analytical solution
$\frac{dP}{dt} = -k_1 P$	$P = P_0 e^{-k_1 t}$
$\frac{dM}{dt} = k_1 P - k_2 M$	$M = \frac{k_1 P_0}{(k_2 - k_1)} \left(e^{-k_1 t} - e^{-k_2 t}\right)$
$\frac{dDP}{dt} = k_2 M$	$DP = \frac{P_0}{(k_2 - k_1)} \left[k_2 \left(1 - e^{-k_1 t}\right) - k_1 \left(1 - e^{-k_2 t}\right)\right]$

Table 1.
Mass balance equations and analytical solutions for the generalized Saeman's model.

into degradation products (DP) (including, HMF from glucose or galactose, and furfural from xylose or arabinose) [26]. When mass balances of these species are made in batch systems and kinetic expressions are introduced, it obtains the ordinary differential equations and respective analytical solutions presented in **Table 1**, for initial conditions $P(0) = P_0$, $M(0) = 0$, and $DP(0) = 0$ [26].

In Eq. (3), k_1 and k_2 represent the kinetic constants for the formation and decomposition first-order reactions of the monomer, respectively, which are temperature dependent (T) according to the well-known Arrhenius law, given in the generalized form by Eq. (4) [26]:

$$k_i = A_i e^{-Ea_i/(R.T)} \quad ; \quad i = 1, 2 \tag{4}$$

In Eq. (4), k_i is the first-order kinetic constant, A_i is the pre-exponential factor, Ea_i is the activation energy, all regarding i-reaction, R is the universal gas constant, and T is the absolute temperature.

As autohydrolysis is a hydrolytic process catalyzed by protons, it is interesting to understand how the evolution of proton concentration occurs during the process, allowing to evaluate the acid level effect on the reaction rates [47]. In this sense, a minor modification in Eq. (4) is proposed in the literature aiming to explicit this effect on the kinetic constants, according to Eq. (5) [47]:

$$k_i = A_i e^{-Ea_i/(R.T)} f(t, [H^+]); i = 1, 2 \tag{5}$$

In Eq. (5), the function $f(t, [H^+])$ represents the dynamic behavior of the hydronium ion concentration, which, for autohydrolysis processes, has been determined empirically in the literature to be of the form of Eq. (6), where a and b are empirical constants. Other authors prefer to consider this effect as constant, already implicitly incorporated into the pre-exponential factor, defined in Eq. (4).

$$f(t, [H^+]) = a + bt \tag{6}$$

An important improvement in Saeman's model was to consider the existence of two hemicellulose fractions separated into two distinct regions in the solid phase: a more susceptible or fast fraction (easy to hydrolyze) and a less susceptible or slow fraction (difficult to hydrolyze), resulting in the so-called "two-phase model" [26, 47, 48]. Thus, based on these premises, a more realistic reaction scheme for the hydrolysis of hemicellulose is represented by:

$$\left.\begin{array}{l} \textit{Fast hemicellulose} \xrightarrow{k_f} \\ \textit{Slow hemicellulose} \xrightarrow{k_s} \end{array}\right\} \textit{Monomer}\ (\mathrm{M}) \xrightarrow{k_2} \textit{Degradation Products}\ (\mathrm{DP}) \tag{7}$$

The main situations for which the use of the two-phase model is indicated are the following [26]: (i) limitations on the mass and energy transfer rates (higher for larger particle size and lower water-solid ratios); (ii) existence of distinct hemicellulose fractions with specific reactivities due to the structure, accessibility, and other factors; (iii) occurrence of uronic acid, which has less reactivity than xylose, reducing the solubilization rate of hemicelluloses; (iv) difference in the reaction rate of the fraction combined with lignina; (v) variations in the contact interfacial area between water and hemicelluloses during the reaction.

The fraction of fast hydrolysis is quantified by the parameter α ($0 < \alpha < 1$), which represents the ratio between the mass of this fraction and the total mass of the two fractions. Therefore, the parameter α must be inserted as a multiplicative factor in the second member of Eq. (8) for *DP* in **Table 1** to better describe the monomer formation during the hydrolysis of hemicelluloses [26, 48].

A simple alternative formulation of the two-phase model considers two solid species of hemicelluloses (fast and slow) denoted as $X_i(s)$, which hydrolyze following first-order kinetics ($r_i = k_i\, X_i$) to form a set of soluble products, X(aq), that are susceptible to subsequent hydrolysis or decomposition reactions, as represented in Eq. (8) [47].

$$X_i(s) \xrightarrow{H^+,\, r_i} X(aq) \tag{8}$$

Approaches based on this formulation can be found to describe the formation of oligomeric intermediates once that oligosaccharides are important as intermediates or substrates [47]. Thus, other models assume that there are two fractions of oligomers in the medium, with most studies focusing on xylan-rich biomass as the predominant hemicellulose [45]. Xylan is a polymer with side chains to which arabinose, acetyl groups, and uronic acids are linked. Generally, for mathematical modeling purposes, xylan has a simplified definition, being the total pentose content in the product fractions, without considering the composition of the original xylan and the variations both in the composition of that which remains in the solid phase and in the composition of the xylo-oligomers throughout hydrolysis [45]. From this viewpoint, a detailed mathematical model, accounting for variations in the compositions of xylan (Xy) and xylo-oligomers, was developed based on the following reaction scheme [45]:

$$\left.\begin{matrix} \mathrm{Xy1} \overset{k_1}{\rightarrow} \\ \mathrm{Xy2} \overset{k_2}{\rightarrow} \end{matrix}\right\} \textit{Xylan oligomers}\ (\mathrm{XO}) \left\{\begin{matrix} \overset{k_8}{\rightarrow} \textit{Acetic acid}\ (\mathrm{ACE}) & \\ \overset{k_3}{\rightarrow} \textit{Xylose}\ (\mathrm{XYL}) & \overset{k_4}{\rightarrow} \\ \overset{k_6}{\rightarrow} \textit{Arabinose}\ (\mathrm{ARA}) & \overset{k_7}{\rightarrow} \end{matrix}\right. \left.\begin{matrix} \\ \\ \\ \end{matrix}\right\} \textit{Furfural}\ (\mathrm{FUR}) \overset{k_5}{\rightarrow} \mathrm{DP} \tag{9}$$

According to this scheme, the model assumes that there is no direct formation of monomers (such as xylose, arabinose, and acetic acid), which are obtained exclusively from the oligomers formed by the hydrolysis of the two xylan fractions [45]. The mass balance equations of the model are given by Eqs. (10)–(18), in which C_j represents the concentration of component j defined as a percentage of the initial mass of xylan that was transformed into component j, and the average composition of the oligomers is represented by $x_{XO\text{ - }XYL}$, $x_{XO\text{-}ARA}$, and $x_{XO\text{-}ACE}$, which are the mass fractions of anhydroxylose, anhydroarabinose, and acetyl, respectively, in the oligomers.

$$C_{\text{Xy}} = C_{\text{Xy1}} + C_{\text{Xy2}} \tag{10}$$

$$\frac{dC_{\text{Xy1}}}{dt} = -k_1 C_{\text{Xy1}} \tag{11}$$

$$\frac{dC_{\text{Xy2}}}{dt} = -k_1 C_{\text{Xy2}} \tag{12}$$

$$\frac{dC_{\text{XO}}}{dt} = k_1 C_{\text{Xy1}} + k_2 C_{\text{Xy2}} - \left(k_3 x_{\text{XO-XYL}} + k_6 x_{\text{XO-ARA}} + k_8 x_{\text{XO-ACE}}\right) C_{\text{XO}} \tag{13}$$

$$\frac{dC_{\text{XYL}}}{dt} = k_3 x_{\text{XO-XYL}} C_{\text{XO}} - k_4 C_{\text{XYL}} \tag{14}$$

$$\frac{dC_{\text{ARA}}}{dt} = k_6 x_{\text{XO-ARA}} C_{\text{XO}} - k_7 C_{\text{ARA}} \tag{15}$$

$$\frac{dC_{\text{ACE}}}{dt} = k_8 x_{\text{XO-ACE}} C_{\text{XO}} \tag{16}$$

$$\frac{dC_{\text{FUR}}}{dt} = k_4 C_{\text{XYL}} + k_7 C_{\text{ARA}} - k_5 C_{\text{FUR}} \tag{17}$$

$$\frac{dC_{\text{DP}}}{dt} = k_5 C_{\text{FUR}} \tag{18}$$

Since average composition of the xylo-oligomers ($x_{\text{XO-XYL}}$, $x_{\text{XO-ARA}}$, and $x_{\text{XO-ACE}}$) varies with time, Eq. (13) must be substituted by Eqs. (19)–(22), which are the mass balances for anhydrous xylose, arabinose, and acetyl in the oligomers [45]:

$$C_{\text{XO}} = C_{\text{XO-XYL}} + C_{\text{XO-ARA}} + C_{\text{XO-ACE}} \tag{19}$$

$$\frac{dC_{\text{XO-XYL}}}{dt} = k_1 x_{\text{Xy1-XYL}} C_{\text{Xy1}} + k_2 x_{\text{Xy2-XYL}} C_{\text{Xy2}} - k_3 x_{\text{XO-XYL}} C_{\text{XO}} \tag{20}$$

$$\frac{dC_{\text{XO-ARA}}}{dt} = k_1 x_{\text{Xy1-ARA}} C_{\text{Xy1}} + k_2 x_{\text{Xy2-ARA}} C_{\text{Xy2}} - k_6 x_{\text{XO-ARA}} C_{\text{XO}} \tag{21}$$

$$\frac{dC_{\text{XO-ACE}}}{dt} = k_1 x_{\text{Xy1-ACE}} C_{\text{Xy1}} + k_2 x_{\text{Xy2-ACE}} C_{\text{Xy2}} - k_8 x_{\text{XO-ACE}} C_{\text{XO}} \tag{22}$$

The average composition of the oligomers is calculated as mass fractions, based on the following equations [45]:

$$x_{\text{XO-XYL}} = \frac{C_{\text{XO-XYL}}}{C_{\text{XO-XYL}} + C_{\text{XO-ARA}} + C_{\text{XO-ACE}}} \tag{23}$$

$$x_{\text{XO-ARA}} = \frac{C_{\text{XO-ARA}}}{C_{\text{XO-XYL}} + C_{\text{XO-ARA}} + C_{\text{XO-ACE}}} \tag{24}$$

$$x_{\text{XO-ACE}} = \frac{C_{\text{XO-ACE}}}{C_{\text{XO-XYL}} + C_{\text{XO-ARA}} + C_{\text{XO-ACE}}} \tag{25}$$

Considering that $k_\text{i} = k_\text{i}(T(t))$, the system of equations can be solved by numerical methods since the initial conditions and the temperature temporal profile are known.

Another interesting mathematical model was proposed in the literature to describe a hydrothermal pretreatment process of lignocellulosic biomass in a large-scale continuous thermal reactor, based on a complex conversion mechanism presented as follows [49]:

$$\begin{aligned}
&\textit{Acetyls} \rightarrow \textit{Organic acids}\\
&\textit{Arabinan} \rightarrow \textit{Arabinose} \rightarrow \textit{Furfural}\\
&\textit{Cellulose} \rightarrow \textit{Glucose} \rightarrow 5-\mathrm{HMF}\\
&\textit{Xylan} \rightarrow \textit{Xylo}-\textit{oligomers} \rightarrow \textit{Xylose}\begin{cases}\rightarrow \textit{Furfural}\\ \rightarrow \textit{Other degradation products}\end{cases}\\
&\textit{Carbohydrates}\begin{cases}\textit{Xylo}-\textit{oligomers}\\ \textit{Xylose}\\ \textit{Arabinose}\\ \textit{Glucose}\end{cases} + \begin{cases}\textit{Furfural}\\ 5-\mathrm{HMF}\end{cases} \rightarrow \textit{Pseudo}-\textit{lignin}
\end{aligned} \tag{26}$$

According to this mechanism, cellulose is partially hydrolyzed to glucose, which is subsequently degraded to 5-HMF. Arabinan is completely converted into arabinose. The xylan-to-xylose conversion step has xylo-oligomers as an intermediate product, and the prediction of these compounds is important since they act as very strong inhibitors in biotechnological processes (enzymatic and fermentative) together with xylose. The pentoses (arabinose and xylose), in turn, are further decomposed into furfural and other degradation products. Carbohydrates (xylo-oligomers, xylose, arabinose, and glucose) react with furfural and 5-HMF to form spherical droplets with a structure similar to lignin, also called pseudo-lignin, which can deactivate enzymes in a subsequent process. Furfural and 5-HMF are important compounds to predict due to their inhibition effects on fermentations, while organic acids alter the pH of sequential enzymatic hydrolysis processes.

The reaction rates, relating to the proposed mechanism, are described by the following first-order equations, in which $k_i = k_i(T(t))$ [49]:

- Rate of glucose production rate (r_G): $r_G = k_G C_C(s)$, $C_C(s)$ = cellulose concentration in the solid phase;
- Rate of glucose degradation rate to 5-HMF (r_H): $r_H = k_H C_G$, C_G = glucose concentration in the liquid phase;
- Rate of arabinan hydrolysis to form arabinose (r_A): $r_A = k_A C_A(s)$, $C_A(s)$ = arabinan concentration in the solid phase;
- Rate of xylan degradation to produce xylo-oligomers (r_{XO}): $r_{XO} = k_{XO} C_X(s)$, $C_X(s)$ = xylana concentration in the solid phase;
- Rate of xylo-oligomers decomposition to produce xylose (r_{XYL}): $r_{XYL} = k_{XYL} C_{XO}$, C_{XO} = xylo-oligomers concentration;
- Rate of furfural production from pentoses (r_F): $r_F = k_F C_P$, C_P = concentration of pentoses (arabinose and xylose) in the liquid phase;
- Rate of pseudo-lignin formation from reaction of carbohydrates with furfural and 5-HMF (r_L): $r_L = k_L C_{CBH}\, C_{CBH}(C_F + C_H)$, C_{CBH} = concentration of carbohydrates (xylo-oligomers, xylose, arabinose, and glucose), C_H = concentration of 5-HMF.

3.2 Mathematical modeling based on severity factor

Pretreatment is the main step in biomass fractionation, since the yield and structure of the products obtained strongly depend on the treatment conditions [50]. Thus, it is important to correlate the chemical modifications of the removed compounds and the process operational variables (e.g. temperature and time) in order to determine the ideal conditions for obtaining the target lignocellulosic products [50]. The intensity of pretreatment is expressed by the severity factor, which describes the combined effect of temperature and time on the material, resulting in different pretreatment conditions [51]. By increasing the pretreatment severity factor, the released hexoses and pentoses can still be degraded to furfural and hydroxymethylfurfural, which are degraded into levulinic and formic acid, respectively, and can act as inhibitors in the steps of enzymatic hydrolysis and fermentation [52].

The origin of combining temperature and reaction time into a single factor emerged during the establishment of kinetic models for kraft pulping processes and has been considered the key point of process control schemes [27]. Therefore, the severity factor has been widely used to facilitate the optimization and control of biomass hydrothermal pretreatment processes [53].

Mathematical models have been developed to better understand the kinetics of the autohydrolysis process and to determine optimal operating conditions to increase reaction rates and yields with minimal energy consumption. Specifically, the use of the severity factor allows us to compare different experimental data on the yield of mono- and oligosaccharides, lignin, and pH, as well as predict such variables as a function of a combination of processing temperature and reaction time [54].

In addition, the severity factor is an appropriate tool that can be used to interpret and analyze experimental data, allowing a rapid and comprehensive understanding of the reaction mechanisms involved in the biomass hydrothermal pretreatment processes, and thus contributing to the design of reactors where such processes are carried out [55].

Conceptually, the severity factor, denoted by S_0, is defined by Eqs. (27) and (28) [44, 51, 55]:

$$R_0 = \int_0^t \exp\left(\frac{T(t) - T_{\text{ref}}}{14.75}\right) dt \xrightarrow{\text{constant } T} R_0 = t \exp\left(\frac{T - T_{\text{ref}}}{14.75}\right) \quad (27)$$

$$S_0 = \log_{10}(R_0) \quad (28)$$

In Eq. (28), t is the reaction time (min), T is the temperature (°C), T_{ref} is the reference temperature (T_{ref} = 100°C), and 14.75 is an empirical constant related to the activation energy, assuming pseudo-first-order kinetics.

By inserting Eq. (27) into Eq. (28) and developing the result, Eq. (29) is obtained, which shows that the severity factor has a linear dependence on the temperature:

$$S_0 = \log_{10}(t) + 0.0294T - 2.94 \quad (29)$$

Eq. (29) provides a quantitative measure of how the severity factor is affected by the two main operating variables (temperature and time). Using equations derived from Eq. (29), some important results can be predicted, such as [55]: (i) severity factor drops rapidly as residence time increases; (ii) for a given temperature, there is a corresponding residence time that results in the same severity factor; (iii) for a constant severity factor, a ten-degree increase in temperature reduces the residence time by half.

Although there is a modified equation in the literature that also considers the influence of pH on the severity factor, for hydrothermal processes, temperature and time are the most relevant variables, making the mentioned equation rarely used.

3.3 Mathematical modeling based on statistical experimental design and multiple linear regression theory

Statistical techniques, including Design of Experiments (DOE) and Response Surface Methodology (RSM), have been successfully used for mathematical modeling and optimization of autohydrolysis processes. Commonly investigated independent variables are reaction time, temperature, and sometimes solid-to-liquid ratio, also known as consistency [43, 44, 56].

Depending on the type of experimental design adopted, linear or quadratic models can be built. However, if the objective is to optimize the system, a quadratic model is necessary. In this case, the most usual experimental design is the Composite Central Rotational Design (CCRD), which is a wide experimental design that allows to obtain a more comprehensive knowledge on the response variable behavior [56]. In addition, it provides sufficient experimental data and degrees of freedom to fit a quadratic model on the significant independent variables [43, 44].

By applying a CCRD to investigate the effects of three independent variables on a system response variable, the full second-order polynomial mathematical model corresponding to this design is given by Eq. (30) [57]:

$$Y = b_0 + b_1X_1 + b_{11}X_1^2 + b_2X_2 + b_{22}X_2^2 + b_3X_3 + b_{33}X_3^2 + b_{12}X_1X_2 + b_{13}X_1X_3 + b_{23}X_2X_3 \tag{30}$$

In Eq. (30), Y is a specific response variable of the hydrothermal process, $\boldsymbol{b}$ is the vector of the model coefficients (determined by multiple linear regression), X_i represents the independent variable i (coded level), and X_iX_j is the interaction term between variables i and j.

The statistical analysis of the experimental data obtained in the experimental design will reveal which terms of the mathematical model are statistically significant. In practice, mathematical models do not contain all the terms presented in Eq. (30) due to the lack of statistical significance of some of them.

The presence of quadratic terms in the mathematical model allows that, in a subsequent step, the levels of the independent variables to be determined by RSM, aiming at the optimization of the response variable.

A CCRD was used in conjunction with the RSM to investigate the influence of time and temperature on the glucose and sucrose yields, obtained in an autohydrolysis process of lignocellulosic materials, keeping the solid-liquid ratio fixed, to optimize the extraction of these sugars aiming at the production of ethanol through fermentation [56]. The time was investigated at two levels: 5 and 30 min, while the temperature was investigated at three: 121, 150, and 180°C, to evaluate the effects of this variable in a wide range of practicable autohydrolysis conditions. The Response Surface Methodology (RSM) was used to optimize the autohydrolysis process, finding the highest concentrations of glucose and sucrose for a temperature of 150°C and a reaction time of 15 min.

The production of xylo-oligosaccharides (XOS) through the hydrothermal pretreatment of by-products of the pulp industry (wood residues) was carried out by using a CCRD, followed by the application of RSM for process optimization [44]. The investigated variables were temperature (126.4; 140.0; 160.0; 180.0; 193.6° C), reaction time (9.6; 30.0; 60.0; 90.0; 110.4 min), and consistency (1.6; 5.0; 10.0; 15.0; 18.4%). These values correspond to the levels of the variables in the experiments of the factorial block, in the replicates at the central point, and in the experiments at the axial points, which were established based on data from hydrothermal pretreatment processes for the production of XOS from different agroindustrial residues [44]. The by-product was divided into two fractions: the first, denominated of original by-product (OB), and the second, consisted of the by-product milled and extracted, denominated of extracted-milled by-product (EMB) [44]. The statistical significance of the model coefficients was then analyzed and the final equations describing the production of XOS for the two by-product fractions were [56]:

$$XOS = b_0 + b_1X_1 + b_2X_1^2 + b_3X_2 + b_4X_2^2 + b_5X_3 + b_6X_3^2 \text{ (OB)} \quad (31)$$

$$XOS = b'_0 + b'_1X_1 + b'_2X_1^2 + b'_3X_2 + b'_4X_2^2 \text{ (EMB)} \quad (32)$$

In Eqs. (31) and (32), the values of the coefficients are, respectively:

- b_0 = 42.392, b_1 = 0.1415, b_2 = −13.9786, b_3 = 2.931, b_4 = −8.9016, b_5 = 0.0025, b_6 = −4.5688
- b'_0 = 45.0279, b'_1 = 0.6961, b'_2 = 16.9541, b'_3 = 2.4383, b'_4 = −10.3935

The subsequent step was to determine, by RSM, the values of the independent variables that maximized the production of XOS (optimization), obtaining the data presented in **Table 2** [44].

For the purpose of validating the mathematical models, additional experiments were carried out under the previously optimized conditions and the results obtained were compared to those predicted by the models, as shown in **Table 3**, verifying good agreement [44].

By-product		**Coded values***	
	X_1 (−)	X_2 (−)	X_3 (−)
OB	0.005061	0.164635	0.000272
EMB	0.0213	0.1173	0
		Actual values	
	T (°C)	t (min)	C (%)
OB	161.0	65.0	10.0
EMB	160.4	63.5	10.0

*X_1 = (T-160)/20; X_2 = (t-60)/30; X_3 = (C-10)/5.

Table 2.
Coded and real values of the independent variables that optimize hydrothermal pretreatment of wood by-products aiming to maximize XOS production.

	Independent variables			Xylan conversion into XOS	
By-product	T (°C)	t (min)	C (%)	Predicted (%)	Experimental (%)
OB-OC*	161.0	65.0	10	42.6	42.3
EMB-OC**	160.4	63.3	10	45.1	43.5

**OB-OC: original by-product subjected to hydrothermal pretreatment under optimized conditions. **EMB-OC: extracted-milled by-product subjected to hydrothermal pretreatment under optimized conditions.*

Table 3.
Predicted and experimental values of xylan conversion to XOS under the optimized conditions of hydrothermal pretreatment of wood by-products (data presented are in percentage, i.e. gram of products per 100 grams of xylan from the by-product, dry weight).

By-product	Model equations and coefficients
SB	$XOS = b_0 + b_1X_1 + b_2X_1^2 + b_3X_2 + b_4X_2^2$ $b_0 = 23.99560; b_1 = 1.35649; b_2 = -6.82656; b_3 = 2.72656; b_4 = -2.66513$
SS	$XOS = b'_0 + b'_1X_1 + b'_2X_1^2 + b'_3X_2 + b'_4X_2^2$ $b'_0 = 36.77311; b'_1 = 9.71410; b'_2 = -5.62845; b'_3 = 2.03623; b'_4 = -6.28371$

Table 4.
Equations and coefficients of the mathematical models developed for the hydrothermal pretreatment of sugarcane by-products.

	Independent variables			Xylan conversion into XOS	
By-product	T (°C)	t (min)	C (%)	Predicted (%)	Experimental (%)
SB	161.98	75.34	10.00	24.76	24.75
SS	177.25	64.86	10.00	41.13	45.34

Table 5.
Data regarding the optimization of the hydrothermal pretreatment of sugarcane by-products and validation of the respective mathematical models.

In a very similar study, also using CCRD and RSM, the optimization of hydrothermal pretreatment of two sugarcane by-products, bagasse (SB) and straw (SS), was carried out for the production of XOS [43]. The independent variables were those from the previously reported study, i.e. temperature, reaction time, and consistency, which were also investigated at the same levels as in the other study [44]. The mathematical models developed, as well as the values of their respective coefficients, are presented in **Table 4** [43].

Subsequently, RSM was applied to determine the optimal conditions for the hydrothermal pretreatment. Experiments under such conditions were carried out and the results obtained were compared to the predictions of the mathematical models to validate them, obtaining the data presented in **Table 5**, which show excellent agreement for SB and reasonable for the SS by-product [43].

4. Conclusion

Lignocellulosic biomass is a renewable source of raw material that can be fractionated into intermediate products of industrial interest. Lignocellulosic fractionation is the main objective of a biomass refinery, consisting of separating the main polymers that form biomass (i.e., cellulose, hemicellulose, and lignin) and processing them further into profitable final products.

Hydrothermal pretreatment of lignocellulosic biomass is an economical method for second-generation biorefineries and, therefore, has been commercialized on a large scale, being conducted in large horizontal thermal reactors and pressurized, with hot liquid water or steam as forms of heating. This treatment method is considered simpler and less expensive, which can significantly reduce the costs of production. Autohydrolysis, as well as other pretreatment methods, aims to obtain hemicellulose and its derivatives for the manufacture of materials and chemical products. The successful ability of autohydrolysis processes to improve the selective depolymerization of cellulose and hemicellulose is well known, making it a viable alternative to the conventional acid hydrolysis method. Furthermore, it allows for less generation of unwanted by-products and a significant reduction in operating costs, as the neutralization step could be eliminated.

In developing kinetic models to describe pretreatment processes, the influence of pretreatment conditions, including temperature, chemical concentration, pH, and residence time, as well as the interactions between lignin and carbohydrates on the reaction rates, needs to be taken into account. Regarding temperature, in general, it was found that kinetic models based on the Arrhenius relationship are adequate to accurately describe the effect of this variable on hemicellulose hydrolysis.

Mathematical models have been developed to better understand the kinetics of the autohydrolysis process, as well as to find optimal operating conditions to increase reaction rates and yields with minimal energy consumption. The literature on the kinetics of hemicellulose removal is extensive, however, mathematical models are generally built upon the classical approach used for dilute acid hydrolysis of cellulose. Pretreatment hydrothermal is a very complex reaction system, whose full mechanisms still need to be elucidated. Therefore, understanding such mechanisms through the use of chemical kinetics principles is of great importance for the applications of hemicellulose. Furthermore, a deeper understanding of pretreatment mechanisms is useful in building robust models that are appropriate for a wide range of pretreatment conditions.

The severity factor is a parameter that incorporates all pretreatment conditions, which can be integrated into kinetic or empirical models to minimize the necessary efforts in process optimization. It was shown that the severity factor is an appropriate parameter to describe the residence time and temperature combination that results in a maximum yield of the desired intermediate product. Thus, this parameter quantifies the combined effects of temperature and time on the transformation of biomass, ensuring that the desired modifications in the structure of the biomass are obtained and allowing the development of the reactor design with a certain speed.

Finally, from the cases studied, the usefulness and effectiveness of experimental design tools for data interpretation, mathematical modeling, and optimization of biomass hydrothermal pretreatment processes were successfully demonstrated.

Author details

Samuel Oliveira[1*], Fernando Paz-Cedeno[2], Naila Mori[3], Adilson Gonçalves[3] and Fernando Masarin[1]

1 School of Pharmaceutical Sciences (FCF), Department of Bioprocess Engineering and Biotechnology (DEBB), São Paulo State University (UNESP), Araraquara, SP, Brazil

2 Division of Agriculture, Dale Bumpers College of Agricultural, Food, and Life Sciences, University of Arkansas System, Fayetteville, AR, USA

3 São Paulo State University (UNESP), Bioenergy Research Institute (IPBEN), Rio Claro, SP, Brazil

*Address all correspondence to: samuel.oliveira@unesp.br

References

[1] Limayem A, Ricke SC. Lignocellulosic biomass for bioethanol production: Current perspectives, potential issues and future prospects. Progress in Energy and Combustion Science. 2012;**38**:449-467

[2] Taniguchi M, Takahashi D, Watanabe D, Sakai K, Hoshino K, Kouya T, et al. Effect of steam explosion pretreatment on treatment with *Pleurotus ostreatus* for the enzymatic hydrolysis of rice straw. Journal of Bioscience and Bioengineering. 2010; **110**(4):449-452

[3] Chen CY, Yeh KL, Aisyah R, Lee DJ, Chang JS. Cultivation, photobioreactor design and harvesting of microalgae for biodiesel production: A critical review. Bioresource Technology. 2011;**102**: 71-81

[4] Han G, Deng J, Zhang S, Bicho P, Wu O. Effect of steam explosion treatment on characteristics of wheat straw. Industrial Crops and Products. 2010;**31**(1):28-33

[5] Ibrahim MM, El-Zawawy WK, Abdel-Fattah YR, Soliman NA, Agblevor FA. Comparison of alkaline pulping with steam explosion for glucose production from rice straw. Carbohydrate Polymers. 2011;**83**(2):720-726

[6] Díaz MJ, Cara C, Ruiz E, Pérez-Bonilla M, Castro E. Hydrothermal pre-treatment and enzymatic hydrolysis of sunflower stalks. Fuel. 2011;**90**(11): 3225-3229

[7] Martín C, Galbe M, Wahlbom CF, Hahn-Hägerdal B, Jönsson LJ. Ethanol production from enzymatic hydrolysates of sugarcane bagasse using recombinant xylose utilizing *Saccharomyces cerevisiae*. Enzyme and Microbial Technology. 2011;**31**(3):274-282

[8] Silva VFN, Arruda PV, Felipe MGA, Gonçalves AR, Rocha GJM. Fermentation of cellulosic hydrolysates obtained by enzymatic saccharification of sugarcane bagasse pretreated by hydrothermal processing. Journal of Industrial Microbiology & Biotechnology. 2011;**38**(7):809-817

[9] NREL. National Renewable Energy Laboratory. Golden, Colorado, USA: Biomass Research; 2009. Available from: http://www.nrel.gov/biomass/biorefinery.htm [Accessed: November 11, 2024]

[10] Mosier N, Wyman C, Dale B, Elander R, Holtzapple YYLM, Ladisch M. Features of promising technologies for pretreatment of lignocellulosic biomass. Bioresource Technology. 2005;**96**: 673-686

[11] Balat M, Balat H, Cahide O. Progress in bioethanol processing. Progress Energy and Combustion Science. 2008; **34**:551-573

[12] Agbor VB, Cicek N, Sparling R, Berlin A, Levin DB. Biomass pretreatment: Fundamentals toward application. Biotechnology Advances. 2011;**29**(6):675-685

[13] Gupta VK, Potumarthi R, O'donovan A, Kubicek CP, Sharma GD, Tuohy MG. Chapter 2 – Bioenergy research: An overview on technological developments and bioresources. In: Bioenergy Research: Advances and Applications. Oxford, UK: Elsevier; 2015

[14] Sun Y, Cheng J. Hydrolysis of lignocellulosic material for etanol production: A review. Bioresource Technology. 2002;**83**:96-103

[15] Martinez A, Speranza M, Ruiz-Duenas FJ, Ferreira P, Camarero S, Guillen F, et al. Bio-degradation of lignocellulosics: Microbial, chemical, and enzymatic aspects of the fungal attack of lignin. International Journal of Microbiology. 2005;**8**:8-30

[16] Prasad S, Singh A, Joshi HC. Ethanol as an alternative fuel from agricultural, industrial and urban residues. Resources, Conservation and Recycling. 2007;**50**: 1-39

[17] Sarkar N, Ghosh SK, Bannerjee S, Aikat K. Bioethanol production from agricultural wastes: An overview. Renewable Energy. 2012;**37**:19-27

[18] Olsson L editors. Biofuels. In: Advances in Biochemical Engineering/ Biotechnology. Vol. 108. Springer, Berlin, Heidelberg. 2012. pp. 67-93. DOI: 10.1007/10_2007_064

[19] Taherzadeh MJ, Karimi K. Enzyme-based hydrolysis processes for ethanol from lignocellulosic materials: A review. BioResources. 2007;**2**(4):707-738

[20] Alvira P, Tomás-Pejó E, Ballesteros M, Negro MJ. Pretreatment technologies for an efficient bioethanol production process based on enzymatic hydrolysis: A review. Bioresource Technology. 2010;**101**:4851-4861

[21] Levan SL, Ross RJ, Winandy, JE. Effects of fire retardant chemicals on bending properties of wood at elevated temperatures. 1990. Available from: http://128.104.77.228/documnts/fplrp/fplrp498.pdf [Accessed: November, 2024]

[22] Winandy JE. Effects of fire retardant treatments after 19 months of exposure at 150F. 1995. Available from: http://www.fpl.fs.fed.us/documnts/fplrn/fplrn264.pdf [Accessed: November, 2024]

[23] Bobleter O. Hydrothermal degradation of polymers derived from plants. Progress in Polymer Science. 1994;**19**:797-841

[24] Garrote G, Dominguez H, Parajo JC. Hydrothermal processing of lignocellulosic materials. Holz als Roh- und Werkstoff. 1999;**57**:191-202

[25] Cybulska I, Brudecki G, Lei H. Chapter 4: Hydrothermal pretreatment of lignocelulosic biomass. In: Gu T, editor. Green Biomass Pretreatment for Biofuels Production, SpringerBriefs in Green Chemistry for Sustaintability. Berlin, Germany: Springer; 2013

[26] Ramirez-Jaramillo MT, Fraga ES. Comparative modelling and optimization of different pretreatment technologies for bioethanol production. Centro Azúcar. 2011;**38**(1):58-74

[27] Wang Z, Xu J, Cheng JJ. Modeling biochemical conversion of lignocellulosic materials for sugar production: A review. BioRes. 2011;**6**(4):5282-5306. DOI: 10.15376/biores.6.4.5282-5306

[28] Yuan Q, Liu S, Ma M-G, Ji X-X, Choi S-E, Si C. The kinetics studies on hydrolysis of hemicellulose. Frontiers in Chemistry. 2021;**9**:1-12. DOI: 10.3389/fchem.2021.781291

[29] Paz-Cedeno FR, Solorzano-Chavez EG, Dias LM, Otaviano CA, Bustamante LJA, Monti R, et al. Composition and chemical structure of hemicelluloses and polysaccharides with capability of gel formation. In: Brienzo M, editor. Hemicellulose Biorefinery: A Sustainable Solution for Value Addition to Bio-Based Products and Bioenergy [Internet]. Singapore: Springer Nature Singapore; 2022.

pp. 111-137. Available from:. DOI: 10.1007/978-981-16-3682-0_4

[30] Devi A, Bajar S, Kour H, Kothari R, Pant D, Singh A. Lignocellulosic biomass valorization for bioethanol production: A circular bioeconomy approach. Bioenergy Research [Internet]. 2022; **15**(4):1820-1841. Available from:. DOI: 10.1007/s12155-022-10401-9

[31] Gellerstedt G, Ek M, Henriksson G. Wood Chemistry and Biotechnology. Vol. 12009. Berlin, Germany: De Gruyter;

[32] Balakshin MY, Capanema EA, Chang HM. MWL fraction with a high concentration of lignin-carbohydrate linkages: Isolation and 2D NMR spectroscopic analysis. Holzforschung. 2007;**61**(1):1-7. DOI: 10.1515/HF.2007.00

[33] Hatfield RD, Ralph J, Grabber JH. Cell wall cross-linking by ferulates and diferulates in grasses. Journal of Science of Food and Agriculture. 1999;**79**: 403-407

[34] Ralph J, Lundquist K, Brunow G, Lu F, Kim H, Schatz PF, et al. Lignins: Natural polymers from oxidative coupling of 4-hydroxyphenyl-propanoids. Phytochemistry Reviews. 2004;**3**:29-60

[35] Masarin F, Gurpilhares DB, Baffa DC, Barbosa MH, Carvalho W, Ferraz A, et al. Chemical composition and enzymatic digestibility of sugarcane clones selected for varied lignin content. Biotechnology for Biofuels [Internet]. 2011;**4**(1):55. Available from: http://www.biotechnologyforbiofuels.com/content/4/1/55

[36] Ambye-Jensen M, Balzarotti R, Thomsen ST, Fonseca C, Kádár Z. Combined ensiling and hydrothermal processing as efficient pretreatment of sugarcane bagasse for 2G bioethanol production. Biotechnology for Biofuels [Internet]. 2018;**11**(1):336. DOI: 10.1186/s13068-018-1338-y

[37] Zhang J, Zou D, Singh S, Cheng G. Recent developments in ionic liquid pretreatment of lignocellulosic biomass for enhanced bioconversion. Sustainable Energy & Fuels [Internet]. 2021;**5**(6): 1655-1667. DOI: 10.1039/D0SE01802C

[38] Canilha L, Santos VTO, Rocha GJM, Almeida E Silva JB, Giulietti M, Silva SS, et al. A study on the pretreatment of a sugarcane bagasse sample with dilute sulfuric acid. Journal of Industrial Microbiology & Biotechnology 2011;38 (9):1467-1475.

[39] Ruiz HA, Conrad M, Sun SN, Sanchez A, Rocha GJM, Romaní A, et al. Engineering aspects of hydrothermal pretreatment: From batch to continuous operation, scale-up and pilot reactor under biorefinery concept. Bioresource Technology. 2020;**299**:122685

[40] Zanuso E, Lara-Flores AA, Aguilar DL, Velazquez-Lucio J, Aguilar CN, Rodríguez-Jasso RM, et al. Kinetic modeling, operational conditions, and biorefinery products from hemicellulose: Depolymerization and solubilization during hydrothermal processing. In: Ruiz HA, Hedegaard Thomsen M, Trajano HL, editors. BT - Hydrothermal Processing in Biorefineries: Production of Bioethanol and High Added-Value Compound. Cham: Springer International Publishing; 2017. pp. 141-160. DOI: 10.1007/978-3-319-56457-9_5

[41] Singh A, Rodríguez Jasso RM, Gonzalez-Gloria KD, Rosales M, Belmares Cerda R, Aguilar CN, et al. The enzyme biorefinery platform for advanced biofuels production.

Bioresource Technology Reports [Internet]. 2019;**7**:100257. Available from: https://www.sciencedirect.com/science/article/pii/S2589014X19301471

[42] Chen H, Liu J, Chang X, Chen D, Xue Y, Liu P, et al. A review on the pretreatment of lignocellulose for high-value chemicals. Fuel Processing Technology [Internet]. 2017;**160**: 196-206. Available from: https://www.sciencedirect.com/science/article/pii/S0378382016311766

[43] Dias LM, Neto FSPP, Brienzo M, de Oliveira SC, Masarin F. Experimental design, modeling, and optimization of production of xylooligosaccharides by hydrothermal pretreatment of sugarcane bagasse and straw. Biomass Conversion and Biorefinery [Internet]. 2023;**13**(14): 12777-12794. DOI: 10.1007/s13399-021-02151-z

[44] Neto FSPP, Roldán IUM, Galán JPM, Monti R, de Oliveira SC, Masarin F. Model-based optimization of xylooligosaccharides production by hydrothermal pretreatment of eucalyptus by-product. Industrial Crops and Products. 2020;**154**:Article 112707. ISSN 0926-6690. DOI: 10.1016/j.indcrop.2020.112707

[45] Nabarlatz D, Farriol X, Montane D. Kinetic modeling of the autohydrolysis of lignocellulosic biomass for the production of hemicellulose-derived oligosaccharides. Industrial and Engineering Chemistry Research. 2004; **43**:4124-4131. DOI: 10.1021/ie034238i

[46] Vallejos ME, Felissia FE, Area MC. Hydrothermal treatments applied to agro- and forest-industrial waste to produce high added- value compounds. BioResources. 2017;**12**(1):2058-2080

[47] Kapu NS, Yuan Z, Chang XF, Beatson R, Martinez DM, Trajano HL. Insight into the evolution of the proton concentration during autohydrolysis and dilute-acid hydrolysis of hemicellulose. Biotechnology for Biofuels. 2016;**9**:224. DOI: 10.1186/s13068-016-0619-6

[48] Chandel AK, Antunes FAF, Arruda PV, Milessi TSS, Silva SS, Felipe MGA. Dilute acid hydrolysis of agro-residues for the depolymerization of hemicellulose: State-of-the-art. In: da Silva S, Chandel A, editors. D-Xylitol. Berlin, Heidelberg: Springer; 2012. pp. 39-61. DOI: 10.1007/978-3-642-31887-0_2

[49] Prunescu RM, Blanke M, Jakobsen JG, Sin G. Dynamic modeling and validation of a biomass hydrothermal pretreatment process - a demonstration scale study. AiChe Journal. 2015;**61**(12):4235-4250. DOI: 10.1002/aic.14954

[50] Tunc MS, van Heiningen ARP. Autohydrolysis of mixed southern hardwoods: Effect of p-factor. Nordic Pulp & Paper Research Journal. 2009; **24**(1):46-51. DOI: 10.3183/npprj-2009-24-01-p046-051

[51] Ruiz HA et al. Severity factor kinetic model as a strategic parameter of hydrothermal processing (steam explosion and liquid hot water) for biomass fractionation under biorefinery concept. Bioresource Technology. 2021; **342**:125961. DOI: 10.1016/j.biortech.2021.125961

[52] Souza RBA. Study of hydrothermal pretreatment and enzymatic hydrolysis of sugarcane straw [thesis]. São Carlos: Federal University of São Carlos (UFSCar); 2016

[53] Ziegler-Devin I, Chrusciel L, Brosse N. Steam explosion pretreatment of lignocellulosic biomass: A mini-review of theorical and experimental

approaches. Frontiers in Chemistry. 2021;**9**:1-7. DOI: 10.3389/fchem.2021.705358

[54] Bassani A, Fiorentini C, Garrido GD, Carullo D, Spigno G. Kinetic model of wheat straw autohydrolysis considering heating and cooling phases. Chemical Engineering Transactions. 2021;**87**: 199-204. DOI: 10.3303/CET2187034

[55] Conrad M-J. Design of a continuous autohydrolysis pretreatment process of annual lignocellulose for industrial application [thesis]. Hamburg: Technology University of Hamburg; 2023

[56] Nori I, Khallaki K, Touil S. Production of bioethanol from *Opuntia ficus-indica* using *Saccharomyces cerevisiae* and water kefir following an autohydrolysis pretreatment. Biofuels. 2024:1-14. DOI: 10.1080/17597269.2024.2371663

[57] Vuddaraju SP, Nikku MY, Chaduvula AIR, Dasari VRRK, Donthireddy SRR. Application of statistical experimental designs for the optimization of medium constituents for the production of L-Asparaginase by *Serratia Marcescens*. Journal of Microbial & Biochemical Technology. 2010;**2**(4): 89-94